RESEARCH JOURNEYS TO NET ZERO

This book provides useful insight into how academics from diverse disciplinary backgrounds, such as science, engineering, technology, social science, policy, design, architecture, built environment, business, and management, have been conducting research into how to realise net zero emissions to address climate change.

This book explores the ways in which countries around the world have pledged to achieve net zero emissions through decarbonisation processes. It presents the highest calibre research and impact activities carried out in the UK, Europe, North America, Australia, Asia, and Africa. Such activities include conceptualisation, opportunity identification, specific case studies, demonstration of proof of concepts, provision of evidence, education of the general public, and knowledge transfer to companies. Further to this, the chapters also bring to light personal career journeys to net zero by current and future international research leaders. From this book, readers will gain a full understanding of net zero research via multiple disciplinary pathways, be inspired by personal accounts, and will learn key methodologies, including quantitative and qualitative approaches.

The diversity of authors and topics make the book widely applicable to a range of fields, and it will be of great interest to researchers, students, practitioners, and decision makers working towards the goals of net zero and decarbonisation.

Kyungeun Sung is a Senior Lecturer in the School of Art, Design and Architecture at De Montfort University, UK.

Patrick Isherwood is a Lecturer in Solar Energy at the Centre for Renewable Energy Systems Technology, Loughborough University, UK.

Richie Moalosi is a Professor of Industrial Design at the University of Botswana.

RESEARCH AND TEACHING IN ENVIRONMENTAL STUDIES

This series brings together international educators and researchers working from a variety of perspectives to explore and present best practice for research and teaching in environmental studies.

Given the urgency of environmental problems, our approach to the research and teaching of environmental studies is crucial. Reflecting on examples of success and failure within the field, this collection showcases authors from a diverse range of environmental disciplines including climate change, environmental communication and sustainable development. Lessons learned from interdisciplinary and transdisciplinary research are presented, as well as teaching and classroom methodology for specific countries and disciplines.

Institutionalizing Interdisciplinarity and Transdisciplinarity
Collaboration across Cultures and Communities
Edited by Bianca Vienni Baptista and Julie Thompson Klein

Interdisciplinary Research on Climate and Energy Decision Making
30 Years of Research on Global Change
Edited by M. Granger Morgan

Transformative Sustainability Education
Reimagining Our Future
Elizabeth A. Lange

Poetry and the Global Climate Crisis
Creative Educational Approaches to Complex Challenges
Edited by Amataritsero Ede, Sandra Lee Kleppe, and Angela Sorby

Research Journeys to Net Zero
Current and Future Leaders
Edited by Kyungeun Sung, Patrick Isherwood, and Richie Moalosi

For more information about this series, please visit: www.routledge.com/Research-and-Teaching-in-Environmental-Studies/book-series/RTES

RESEARCH JOURNEYS TO NET ZERO

Current and Future Leaders

Edited by Kyungeun Sung, Patrick Isherwood, and Richie Moalosi

LONDON AND NEW YORK

Designed cover image: Eric Ward © Unsplash

First published 2024
by Routledge
4 Park Square, Milton Park, Abingdon, Oxon OX14 4RN

and by Routledge
605 Third Avenue, New York, NY 10158

Routledge is an imprint of the Taylor & Francis Group, an informa business

British Library Cataloguing-in-Publication Data
A catalogue record for this book is available from the British Library

ISBN: 978-1-032-46210-3 (hbk)
ISBN: 978-1-032-46209-7 (pbk)
ISBN: 978-1-003-38056-6 (ebk)

DOI: 10.4324/9781003380566

Typeset in Optima
by Taylor & Francis Books

CONTENTS

FIGURES

ABOUT THE EDITORS

Kyungeun Sung is a Senior Lecturer in the School of Art, Design and Architecture at De Montfort University, UK. Her research is broadly concerned with design and sustainability, focusing on upcycling, circular economy, and net zero. She has research interests and expertise in sustainable design, craft, production, business, supply chains, behaviour, consumption, and lifestyles. Her research uses design as a tool for new product development, service innovation, behavioural change, scaling up niche sustainable practices/behaviours, and transitioning to circular economy and net zero. She has published over 30 peer-reviewed journal articles, conference proceedings, and book chapters in her field with high-impact journals and high-quality publishers. She has edited *State-of-the-Art Upcycling Research and Practice*. She has been appointed to the UKRI (UK Research and Innovation) Interdisciplinary Assessment College and AHRC (Arts and Humanities Research Council) Peer Review College. She has reviewed funding applications in her area for AHRC, EPSRC (Engineering and Physical Sciences Research Council), Italian Ministry for University and Research, and Mauritius Research and Innovation Council. She has reviewed over 50 research papers for international journals and conferences in her area. She is currently the Principal Investigator of the AHRC-funded International Upcycling Research Network (AH/W007134/1).

Patrick Isherwood is a Lecturer in Solar Energy at the Centre for Renewable Energy Systems Technology at Loughborough University, UK. His research interests include developing sustainable materials, devices and processing techniques for solar energy conversion and storage, solar resource modelling, and thermodynamic modelling of solar collectors. He is particularly interested in developing environmentally benign techniques for materials synthesis and end-of-life device processing. He has authored more than 20 peer-reviewed papers and conference proceedings and is frequently invited to review articles for journals in solar energy and related fields.

Richie Moalosi is a Professor of Industrial Design at the Department of Industrial Design and Technology at the University of Botswana. He is also the founding Director of the University of Botswana Innovation Centre. He coordinates the University of Botswana Design for Sustainability and Social Innovation Lab. He has over 20 years of teaching experience at the university level. He works with small micro-enterprises, start-ups, community-based organisations, and the creative industries to develop and add value to their products, brands, and service innovation capabilities, as these are the next engine of the creative economy in Africa and other emerging economies. His specialisation and research interest areas include the following: design and culture, design research, design education, design for sustainability, social innovation, and additive manufacturing. He has also published extensively in international peer-reviewed journals, contributed book chapters, published books, and presented at many international peer-reviewed conferences. He is an editorial board member of several international peer-reviewed journals. He is a member of the Institute of Engineering Designers, Pan Afrikan Design Institute, and Africa Design (Chapter of the Design Society). He is the Co-principal Investigator of the AHRC-funded International Upcycling Research Network (AH/W007134/1).

CONTRIBUTORS

Walter Chipambwa is a Lecturer in the Department of Clothing and Textile Technology, School of Art and Design at Chinhoyi University of Technology in Zimbabwe. He has been a lecturer for the past 11 years with interests in product design, computer-aided design, innovation, sustainable design, creativity, and design education and management.

Patrick Dichabeng is a Lecturer of Product design at the University of Botswana. He holds a Master's in Integrated Product Design. He specialises in computer-aided product design, rapid prototyping, design automation, human factors, and acceptance. He is currently doing his PhD on human factors of highly automated vehicles.

Rafael Gomez, Fellow of the Design Institute of Australia, is Associate Professor in Industrial Design at Queensland University of Technology and Founder of the BMW Group + QUT Design Academy. The academy is founded on internships, research, and special projects working with BMW Group in Germany on world-first innovations across digitalisation and emerging technologies. Dr Gomez's expertise is in emotional experience, wearable tech, mobility, transportation, and qualitative research methods. In addition, Dr Gomez has over 20 years industry experience as a designer for small, medium and large enterprises in aviation, construction, medical, health, food, government and consumer electronics industries.

Luksa Kujovic is a PhD candidate at the Centre for Renewable Energy Systems Technology, Loughborough University. Luksa obtained his BSc in Electrical Engineering and MSc in Renewable Energy Systems Technology.

His research endeavours to improve the performance of thin film cadmium telluride (CdTe) solar cells by developing stable buffer layer materials.

Gustav Markkula is a Professor of Applied Behaviour Modelling at the University of Leeds. Gustav Markkula received a PhD in mathematical modelling of driver behaviour to support virtual testing of automotive safety systems. He is an engineer by training. He applies quantitative methods and models to studying human behaviour and cognition in road traffic.

Natasha Merat is an experimental psychologist and research group leader of the Human Factors and Safety Group in the Institute for Transport Studies at the University of Leeds. She also leads the Automation theme at Leeds and is responsible for the strategic direction of research conducted at Virtuosity. Her main research interests are understanding road users' interaction with new technologies.

Sophia N. Njeru is an educator and researcher in fashion design at Kirinyaga University, Kenya. Njeru is avid about sustainable fashion research and practice. She presents research findings at scientific conferences. Professional memberships include the International Federation for Home Economics (IFHE) and the Learning Network on Sustainable Energy Systems (LeNSes).

Yaone Rapitsenyane is the Head of the Department of Industrial Design and Technology at the University of Botswana. He is also a service designer and a lecturer in sustainable design. His research interests include developing sustainable business models for SMEs and Sustainable Product-Service System curricula for African universities.

Rebecca Roberts is Chief Operating Officer and co-founder of start-up EV8 Technologies. EV8 builds software solutions that support electric vehicle (EV) adoption, electricity grid integration, and energy optimisation. Rebecca's primary role is to lead innovation and development for the company, particularly focusing on vehicle-to-grid and the commercial opportunities in this space.

Oanthata Jester Sealetsa is a Senior Lecturer in the Department of Industrial Design and Technology at the University of Botswana. His research interests include ergonomics and design, occupational health and safety, human-centred design, and Technology Transfer. He is a member of the Ergonomics Society of South Africa.

Prabhu Selvaraj is a Research Associate in advanced building Façade Design at the Centre for Renewable Energy Systems Technology, Loughborough University. His research primarily focuses on three areas: developing vacuum glazing prototypes, building and characterising transpired inverted absorber collector-based façade components, and developing façade system component models.

Hyunjae Daniel Shin is an Assistant Professor in the Department of Integrated Design at Yonsei University, South Korea. He is currently a director of the Design for Sustainable Urban Living Lab. His research focuses on advancing knowledge in the field of Design for Sustainable Behaviour, exploring the impact of design-led intervention on tackling many social issues such as health, safety, and energy.

Jagdeep Singh is an Associate Professor in Environmental Science at the Centre for Environmental and Climate Science, Lund University, Sweden. Dr Singh has advanced the field of sustainable production and consumption in emerging areas such as circular economy, reuse, product repair and upcycling, the sharing economy, and sustainable business innovation.

Abhishek Tiwary is an Associate Professor in the School of Engineering and Sustainable Development at De Montfort University, UK. He is a recipient of Marie Curie Fellowship and a Royal Academy of Engineering Industrial Fellowship to pursue interdisciplinary research at the energy–environment nexus. His current research involves system-scale sustainability assessment of technology innovations, in the UK and international settings.

Garrath T. Wilson is a Senior Lecturer in Experience Design and Lead of the Responsible Design Research Group at Loughborough University's School of Design & Creative Arts. Combining research expertise with design practice, Dr Wilson has written, consulted, and presented internationally on sustainability and designing for the circular economy.

Sun-Jin Yun is a Professor and Dean of the Graduate School of Environmental Studies at Seoul National University in the Republic of Korea. She was the first co-chair of the 2050 Carbon Neutrality and Green Growth Commission of the Republic of Korea. Her major research areas include climate, energy, and environmental policies; environmental and energy transition movements; environmental awareness; environmental, energy, and climate governance; management of common-pool resources; and environmental education. She promotes sustainable development through vigorous research, lecturing, and social activities.

FOREWORD

In the ongoing quest to address the pressing challenges of our time, few endeavours are as paramount as pursuing a sustainable and environmentally responsible future. As a guiding principle on this journey, Net Zero has come to signify a promising commitment to strike a delicate balance between human activity and the fragile ecosystems that support our planet. Achieving Net Zero is a goal that cuts across sectors, industries, and disciplines because of its substantial consequences for ecological equilibrium and climate change mitigation.

Research Journeys to Net Zero: Current and Future Leaders is a monument to the global research community's unwavering commitment and creativity in these pressing difficulties. This book takes us on a unique journey through the personal experiences and Net Zero research of educators, academics, practitioners, and activists from various professions and continents who are at the forefront of the field. They have laid the groundwork for dramatic change through their combined intellect, unrelenting curiosity, and unwavering dedication to a sustainable future.

This collection of journeys is an eye-opener on how external factors, often caused by human activities, have triggered a change in individuals' perception of the potential impact of their work and ignited a passion for contributing to a real transformation in how we live on this planet. Remarkably, this passion is consistent across disciplines, countries, and age groups.

This book provides hope that Net Zero is not an empty promise and its implementation can empower people to embrace new directions. It highlights the pioneers forging new paths, posing essential questions, and proposing ground-breaking ideas and strategies. The necessity to adopt strategies and tools from other research domains is discussed, along with unique approaches and cutting-edge assessment frameworks. For instance, they offer methods of changing design, the built environment, and engineering from extractive and exploitative to those that add positive ecological and social value. The contributions in this book span a wide range of fields, reflecting the interdisciplinary character of the Net Zero challenge, including renewable energy, transportation, engineering, several design professions, carbon capture, and the built environment.

This book is an invaluable resource for individuals trying to understand the complexity of Net Zero research because it offers insights into the current state of knowledge and also envisioning future horizons. These stories help us better understand the pressing need to move toward a sustainable society (circular economy) and the enormous capacity of human intellect to accomplish the same.

I commend the authors and contributors who poured their intellect, passion, and dedication into this magnificent book. The insights and stories shared in this book will inspire others to embark on the critical journey to Net Zero, creating a better and more sustainable future for future generations.

Raffaella Villa
Professor of Environmental Bioengineering
Director, Net Zero theme
De Montfort University

ACKNOWLEDGEMENTS

We would like to thank all the authors who have provided contributions, without which we would not have been able to produce this book. This project was supported by the De Montfort University Vice Chancellor's Future Research Leaders Programme Development Fund and the UKRI (UK Research and Innovation) AHRC (Arts and Humanities Research Council) Research Networking Grant for the International Upcycling Research Network (AH/W007134/1). We are grateful for the generous Research Award at De Montfort University, which allowed us the time to devote to this book. Finally, our deepest appreciation goes to our families and friends for their love and support throughout the process.

ABOUT THIS BOOK

Addressing climate change is one of the most pressing challenges facing the world today, and many countries, including the UK, have pledged to achieve net zero emissions by decarbonisation processes. Academics in numerous disciplines have been conducting research and impact activities on this important topic in terms of, for example, conceptualisation, opportunity identification, case studies, demonstration of proof of concepts, provision of evidence, education of the general public, and knowledge transfer to companies. Besides such activities, informing and inspiring future researchers is a crucial role of current and future research leaders in this area to achieve net zero for the long term. This book presents the highest calibre research and impact activities as well as personal journeys in/to net zero as autoethnographic descriptions by a selection of current and future research leaders. The main benefits of reading this book are, firstly, to gain an understanding of the global net zero research horizon with wide-ranging research avenues from diverse disciplinary backgrounds such as science, engineering, technology, social science, policy, design, business, and management; and secondly, to be inspired by the different academic/professional pathways and personal journeys of the international research leaders working on net zero. The majority of the book chapters came from the Net Zero Conference 2022, which was held on 24 June 2022 at De Montfort University (DMU) in Leicester, UK, as a joint effort between DMU and Loughborough University, as well as contributions by the network members of the AHRC-funded International Upcycling Research Network (AH/W007134/1). The diversity of the authors and topics is applicable and relevant to a wide range of fields and readers.

This book is highly relevant to many countries and regions that aim to achieve carbon neutrality or decarbonisation, such as the UK, Europe, and North America (and potentially the rest of the world) based on the global common goal of achieving net zero and similar academic practices across countries. This book is for undergraduate (UG), postgraduate taught (PGT)/ master's, and postgraduate research (PGR)/PhD students, early career academic researchers/academics, and practitioners and decision makers in government and business spheres who are interested in or working on net zero in a variety of disciplines and sectors across the world. You may want to use this book for:

i   information on important past, current, and future research areas and avenues in/to net zero from diverse disciplinary backgrounds;
ii general opportunities and difficulties faced by researchers working on net zero and how to overcome obstacles;
iii insight into how to progress a research/academic career through different pathways; and
iv inspiration on how to make a positive change in the world with research and impact activities in/to net zero.

On the one hand, this book can be an excellent introductory text for general sustainability, which encourages beginners to join the net zero movement. On the other hand, this book could offer a valuable guide for other researchers doing similar work.

ABBREVIATIONS

3D	three-dimensional
6D	six degrees
AICRIE	Annual International Conference on Research and Innovation in Education
AM	air mass
ASHP	air source heat pump
a-Si	amorphous silicon
AHRC	Arts and Humanities Research Council
AI	artificial intelligence
ALTs	accelerated lifetime tests
AMRs	autonomous mobile robots
AR	augmented reality
ATP	adenosine triphosphate
AZO	aluminium-doped zinc oxide
BETA-EBM	Bio-energy Technology Assessment – Environmental Burden Minimisation
BEV	battery electric vehicle
BICPV	building-integrated concentrating photovoltaics
BIPV	building-integrated photovoltaics
BIPW	building-integrated photovoltaic windows
BSI	British Standards Institution
CBD	chemical bath deposition
CCC	carbon, control and comfort
CCCN	Citizens' Council for 2050 Carbon Neutrality
CCT	correlated colour temperature
CCUS	carbon capture, utilisation and storage
CdO	cadmium oxide

CdS	cadmium sulphide
CdTe	cadmium telluride
CdZnS	cadmium zinc sulphide
CEA	Center for Energy Alternatives
CEC	Centre for Environmental and Climate Science
CFCs	chlorofluorocarbons
CH4	methane
CHP	combined heat and power
CIE-MAP	Centre for Industrial Energy, Material and Products
CIGS	copper indium gallium diselenide
CLEVER	closed loop emotionally valuable e-waste recovery
CNC	Carbon Neutrality Commission
CNS	carbon neutrality scenario
CO2	carbon dioxide
COMPON	Comparative Policy Network
COP	coefficient of performance
COP26	26th Conference of the Parties to the United Nations Framework Convention on Climate Change
COP27	27th Conference of the Parties to the United Nations Framework Convention on Climate Change
CORE	Creative Outreach for Resource Efficiency
CPUT	Cape Peninsula University of Technology
CREST	Centre for Renewable Energy Systems Technology
CRI	colour rendering index
c-Si	crystalline silicon
CZTS	copper zinc tin sulphide
DATTARC	Design and Technologies Teaching and Research Conference
DERs	distributed energy resources
DfS	design for sustainability
DHW	domestic hot water
DIY	do it yourself
DMU	De Montfort University
DN	distribution network
DRE	distributed renewable energy
DRS	Design Research Society
DSO	distribution system operator
DSSC	dye-sensitised solar cell
FMA	Environmental Management Agency
EngD	engineering doctorate
ENHANCE	Enabling Hybrid Autonomous Nonconventional System for Clean Indoor Environment
EOL	end-of-life
EPSRC	Engineering and Physical Sciences Research Council
ESA	European Space Agency

EV	electric vehicle
EVA	ethylene vinyl acetate
FCEV	fuel cell electric vehicle
FDfS	fashion design for sustainability
FTO	fluorine-doped tin oxide
GFED	Global Fire Emissions Database
GHGs	greenhouse gases
GJHSS	*Global Journal of Human Social Sciences*
GMZO	gallium-doped magnesium zinc oxide
HEIF	Higher Education Innovation Funding
HEV	hybrid electric vehicles
HFCs	hydrofluorocarbons
HVAC	heating, ventilation, and air-conditioning
ICE	internal combustion engine
ICED	International Conference on Engineering Design
ICRSC	International Online Conference on Reuse, Recycling, Upcycling, Sustainable Waste Management, and Circular Economy
IFHE	International Federation for Home Economics
IGCAR	Indira Gandhi Centre for Atomic Research
IIIEE	International Institute for Industrial Environmental Economics
IIT	Indian Institute of Technology
IJHE	International Journal for Home Economics
IPCC	Intergovernmental Panel on Climate Change
ISCF	Industrial Strategy Challenge Fund
ISO	International Organization for Standardization
ISWA	International Solid Waste Associate
IT	information technology
ITO	indium tin oxide
IURN	International Upcycling Research Network
KEBS	Kenya Bureau of Standards
KFEM	Korea Federation for Environmental Movements
KIDeC	Kampala International Design Conference
KNMI	Koninklijk Netherlands Meteorologisch Instituut (Royal Netherlands Meteorological Institute)
KTU	Korean Teachers Union
K-POPONS	Korean Professors' Organization for a POst Nuclear energy Society
LCA	life cycle assessment
LCB	Leicester Creative Businesses
LCPV	low-concentration photovoltaic
LEEDR	low effort energy demand reduction
LeNS	Learning Network on Sustainability
LeNSes	Learning Network for Sustainable Energy Systems
mc-Si	monocrystalline silicon

MZO	magnesium zinc oxide
M.Tech	Master of Technology
N2O	nitrous oxide
NDC	nationally determined contribution
NERC	Natural Environment Research Council
NF3	nitrogen trifluoride
NPPs	nuclear power plants
NSG	nippon sheet glass
NTU	Nottingham Trent University
OECD	Organisation for Economic Cooperation and Development
OED	Oxford English Dictionary
OLNPP	one less nuclear power plant
ORWARE	ORganic WAste REsearch
PAR	participatory action research
pc-Si	polycrystalline silicon
PEOH	people exposed to occupational hazards
PFCs	perfluorocarbons
PHEV	plug-in hybrid electric vehicle
PKGWP	Pan-Korea Grand Waterway Project
PLATE	product lifetimes and the environment
PLWD	people living with disabilities
POMAC	Professors' Organization for Movement Against Grand Korea Canal
PPFTG	perpetual plastic for food to go
PV	photovoltaics
PVT	photovoltaic thermal
QUT	Queensland University of Technology
RA	research associate
RAE	Royal Academy of Engineering
RF	radio-frequency
SAE	Society of Automotive Engineers
SAVs	shared automated vehicles
SAZ	Standards Association of Zimbabwe
SDGs	Sustainable Development Goals
SECURE	SElf Conserving URban Environment
SF6	sulphur hexafluoride
SHC	second-hand clothes
SMEs	small and medium-sized enterprises
SnO2	tin oxide
SNP	special needs persons
S.PSS	sustainable product-service system
SORDI	synthetic object recognition dataset for industries
SSPP	smart sustainable plastic packaging
STC	solar thermal collector

SUE	sustainable urban environment
SusSIG	Sustainability Special Interest Group
TCO	transparent conducting oxide
TERI	The Energy and Resources Institute
TN	transmission network
TSO	transmission system operator
UAE	United Arab Emirates
UKRI	UK Research and Innovation
UN	United Nations
UNFCCC	United Nations Framework Convention on Climate Change
UTAUT2	Unified Theory of Acceptance and Use of Technology
UV	ultraviolet
UX	user experience
V2G	vehicle-to-grid
Voc	open circuit voltage
VPP	virtual power plant
VR	virtual reality
WEEE	waste electrical and electronic equipment
ZnO	zinc oxide

INTRODUCTION

Current and future leaders of net zero research and their research journeys

Kyungeun Sung, Patrick Isherwood and Richie Moalosi

The UK government's Net Zero Strategy explains what net zero is and why we need to act:

> If we fail to limit global warming to 1.5°C above pre-industrial levels, the floods and fires we have seen around the world […] will get more frequent and more fierce, crops will be more likely to fail, and sea levels will rise driving mass migration as millions are forced from their homes. Above 1.5°C, we risk reaching climate tipping points like the melting of arctic permafrost – releasing millennia of stored greenhouse gases – meaning we could lose control of our climate for good. But the good news is that there is, still, a path to avoid catastrophic climate change. […] by the middle of this century, the world has to reduce emissions to as close to zero as possible […] If we can achieve this, global emissions of greenhouse gases will be 'net zero'.
>
> *(HM Government, 2021, p.14)*

To further explain the context, the Paris Agreement (UNFCCC, 2015a) was a landmark deal that began a shift towards net zero. Article 4 of the Paris Agreement, building on the 5th Assessment Report of the Intergovernmental Panel on Climate Change (IPCC, 2013) and the United Nations Framework Convention on Climate Change (UNFCCC) Structured Expert Dialogue (UNFCCC, 2015b), states,

> In order to achieve the long-term temperature goal set out in Article 2 [holding the increase in the global average temperature to well below 2°C above pre-industrial levels and pursuing efforts to limit the temperature increase to 1.5°C], Parties aim to […] achieve a balance between

DOI: 10.4324/9781003380566-1

anthropogenic emissions by sources and removals by sinks of greenhouse gases (GHGs) in the second half of this century.

(UNFCC, 2015a, p.4)

In the 26th Conference of the Parties to the United Nations Framework Convention on Climate Change (COP26) in Glasgow in 2021, the Glasgow Climate Pact, recalling Article 2 of the Paris Agreement (see above), acknowledged that "climate change is a common concern of humankind" (UNFCCC, 2021, p.1) and recognised that:

limiting global warming to 1.5°C requires rapid, deep and sustained reductions in global greenhouse gas emissions, including *reducing global carbon dioxide emissions by 45 per cent by 2030 relative to the 2010 level and to net zero around mid-century*, as well as deep reductions in other greenhouse gases.

(ibid., p.3, emphasis added)

During COP26, countries agreed to a provision calling for a phase-down of coal power and a phase-out of inefficient fossil fuel subsidies. Many significant deals and announcements were made, such as 130 countries' Global Methane Pledge aiming to limit methane emissions by 30 per cent by 2030, and over 30 countries' determination (including six major vehicle manufacturers and other actors) for all new car and van sales to be zero-emission vehicles by 2040 globally and 2035 in leading markets (UN, 2021).

In the 27th Conference of the Parties to the United Nations Framework Convention on Climate Change (COP27), the Sharm el-Sheikh Implementation Plan notes "the importance of transition to sustainable lifestyles and sustainable patterns of consumption and production for efforts to address climate change [... and] the importance of pursuing an approach to education that promotes a shift in lifestyles while fostering patterns of development and sustainability based on care, community and cooperation" (UNFCCC, 2022, p.1), recognises "the importance of the best available science for effective climate action and policymaking" (ibid., p.2), and stresses "the importance of enhancing a clean energy mix, including low-emission and renewable energy, at all levels as part of diversifying energy mixes and systems, in line with national circumstances and recognising the need for support towards just transitions" (ibid., p.3).

Achieving net zero requires operationalisation in varied socio-political and economic spheres as there are many potential pitfalls and factors to be considered (e.g., behavioural, economic, ethical, legal, political, social, technological) (Fankhauser et al., 2022). It is one of the most pressing and complex challenges facing the world today. It requires research and actions for meaningful change and significant impact across different industry sectors and socio-political and economic spheres. Recognising the challenge, many scientists, researchers, and academics in varied disciplines have conducted research on this important topic in order to understand the challenge in different contexts better, identify promising opportunities and best practices for interventions, demonstrate proof of concepts, provide evidence for better policy making, transfer knowledge to companies, and educate the general public. Besides such essential research activities, informing and inspiring future researchers is also a crucial role of current and future research leaders in this area to achieve net zero for the long term.

That is why we organised the Net Zero Conference 2022 at De Montfort University (DMU) in Leicester, UK, in June 2022, focusing on current and future research leaders' research journeys in/to net zero. The conference was a joint event between DMU and Loughborough University supported by the DMU Vice Chancellor's Future Research Leaders Programme Development Fund and the UKRI (UK Research and Innovation) AHRC (Arts and Humanities Research Council) Research Networking Grant for the International Upcycling Research Network (IURN). Most of this edited book's contributions are from speakers at the Net Zero Conference 2022 and IURN members. The contributions are the highest calibre research and impact activities and personal journeys in/to net zero as autoethnographic descriptions by the current and future research leaders. Five themes emerged from the contributions, which form the five parts of this book: Part I Science and Engineering, Part II Design and Innovation, Part III Energy Sector, Part IV Transportation Sector, and Part V Fashion and Service Sectors.

In Part I Science and Engineering, Prof Sun-Jin Yun at Seoul National University in South Korea opens the first chapter and talks about her life as a teacher, professor, researcher, practitioner, activist, and institutional participant working on environment, energy and climate change issues. Dr Yun has been deeply involved in developing Korea's 2050 carbon neutrality scenario and the 2030 nationally determined contribution as the first co-chair of Korea's 2050 Carbon Neutrality Commission. She believes that developing the 2050 carbon neutrality scenario is neither sufficient nor complete but is the first step toward building a climate-safe society. In Chapter 2, Dr Jagdeep Singh, in the Centre for Environmental and Climate Science at Lund University in Sweden, explains the role of disciplinary, interdisciplinary, and transdisciplinary education in sustainability science using his own experiences from his professional journey. Dr Singh argues that understanding and addressing complex sustainability challenges, including net zero, requires applying novel approaches, devising innovative assessment frameworks, and adopting methods and tools across research fields and disciplines. He believes his multidisciplinary educational background and varied work experiences across disciplines have enriched his journey, allowing metacognition and deep learning about sustainability challenges. In the last chapter in this part (Chapter 3), Dr Abhishek Tiwary from De Montfort University in the UK talks about the role of evolutionary science/engineering in shaping his academic and research journey into net zero and describes his research over the past two decades, enriched from the interdisciplinary remits of net zero challenges. Dr Tiwary's latest research has focused on integrating sustainability into engineering projects, addressing the dilemma of achieving sustainable net zero at the energy–environment nexus.

In Part II Design and Innovation, Dr Rafael Gomez at Queensland University of Technology (QUT) in Australia presents an overview of the unique partnership between BMW Group and QUT Design Academy, highlighting their leadership and commitment to ambitious sustainability targets. Dr Gomez introduces BMW Group's six megatrends, including humanisation and emission-free energy as the foundation of their vision for the future, and then explains how design aligns closely with these megatrends with its human-centred, creative, and sustainable approaches. He showcases cutting-edge projects utilising digital tools to inspire other researchers to engage with leading corporations for a sustainable future. The successful partnership between BMW and QUT Design Academy demonstrates that design drives sustainability and innovation. In Chapter 5, Dr Garrath Wilson, Lead of the Responsible Design Research Group at Loughborough University in the UK, provides a reflection on his career journey, from a designer to early career design researcher through to Principal Investigator of a substantial, sustainable plastic packaging research project, and how his journey has been shaped by both global and personal challenges. Dr Wilson also discusses the value of allowing oneself, as an early career researcher, the freedom to pivot context and collaborators and to recognise the positive side of a bit of randomness. In Chapter 6, Dr Hyunjae Shin, the Director of Design for Sustainable Urban Living Lab at Yonsei University in South Korea, describes his path to finding his passion for designing human-powered products. Dr Shin discusses his experiences as a researcher exploring the benefits of human-powered products as a viable choice for sustainable living. He also discusses his current work in the field of Design for Sustainable Behaviour. In Chapter 7, Dr Kyungeun Sung at De Montfort University in the UK describes her research journey on sustainable production and consumption by upcycling towards net zero from the start of her PhD in 2013 until 2023. About ten years of her research in and around upcycling covers sustainable art, design, craft, production, business, supply chains, behaviour, consumption, and lifestyles.

In Part III Energy Sector, three researchers at different stages of their careers in CREST (Centre for Renewable Energy Systems Technology) at Loughborough University in the UK describe their cutting-edge research and academic and personal journeys. In Chapter 8, Dr Patrick Isherwood talks about his unusual and convoluted route into photovoltaics research and uses this narrative to discuss a range of photovoltaic (PV) and PV-related technologies. Dr Isherwood discusses thermodynamics and loss mechanisms in solar cells, the problems associated with PV panel recycling, and the need to develop PV panels with minimal to no polymeric constituents. He argues that further research and development should focus on understanding the physical limits to photovoltaic energy conversion and the means of fully recycling PV modules and devices as they reach their end-of-life. In Chapter 9, Dr Prabhakaran Selvaraj discusses the importance of semi-transparent solar panels for windows to achieve zero energy buildings, increase the aesthetic value of the built environment, and the challenges in scaling up these solar cells. He also reviews the current state-of-the-art science and technology for window glazing systems. He discusses future research in vacuum glazing and the possibilities for improving the thermal insulating performance of buildings to achieve the net zero goal. In Chapter 10, Luksa Kujovic talks about his research journey, beginning with a focus on maximising the integration of distributed energy resources into the distribution network while managing the increased uncertainties in the system. He uses accelerated lifetime tests to describe his investigation into the environmental stability of various n-type buffer layers for thin film cadmium telluride (CdTe) solar cells. His findings highlight zinc oxide (ZnO) as the most stable buffer layer, paving the way for further research and optimising alternative buffer layers in thin-film CdTe solar cells.

In Part IV Transportation Sector, Dr Rebecca Roberts, Chief Operating Officer, and Co-Founder of EV8 Technologies (electric vehicle adoption software solution start-up) in the UK, describes her non-linear journey from architecture student to systems engineer via renewable energy systems research in Chapter 11. Dr Roberts talks about her interest in practical and commercial applications of technology research and discusses her research focusing on the commercial opportunities for a variety of renewable and adjacent technologies – photovoltaics, heat pumps, solar thermal, electric vehicles as battery storage and vehicle-to-grid – with a core focus on the applications and critical considerations for vehicle-to-grid, specifically the impact of user behaviour on the business case. In Chapter 12, three academics at the University of Botswana (Prof Richie Moalosi, Dr Yaone Rapitsenyane, and Oanthata Jester Sealetsa) describe a move from a fossil-energy-dominated transport industry towards an equitable, clean, and sustainable mobility ecosystem built in local micro-factories in Africa, presenting an opportunity for decarbonisation. They discuss the case studies in Africa regarding the role of electric vehicle infrastructure, electricity accessibility, barriers, and opportunities to transition to net zero in the transport sector. They argue that it will be impossible to reach net zero emissions in transport by 2040 in Africa due to barriers such as lack of incentives, finance, unavailability of public charging infrastructure, skilled labour force, and high import taxes, and call for African governments to invest in new infrastructure to produce clean energy and electric mobility. Chapter 13 presents the collaborative research on exploring user acceptance of shared automated vehicles (SAVs) between three academics from the University of Botswana (Patrick Dichabeng) and the University of Leeds (Prof Natasha Mera and

Prof Gustav Markkula). They describe their research, which explores the attitudes, perceptions, and preferences influencing the adoption of SAVs among this crucial user group. They found that service quality, trust, and price value are the three most prominent factors affecting user acceptance of SAVs. The insights from this study could inform the development of targeted strategies and policies to encourage the transition to SAVs, supporting the global push towards sustainable urban mobility and net zero emissions.

In Part V Fashion and Service Sectors, Dr Yaone Rapitsenyane, at the University of Botswana, discusses the integration of Sustainable Product-Service System (S.PSS) and service design for net zero in developing economies in Chapter 14. Dr Rapitsenyane presents his work on integrating sustainable design, product-service systems, and service design to demonstrate the evolution of his research work as a scholar of sustainable design. He explains the tools used to develop and communicate solutions and high-level processes in Botswana. He concludes by proposing future directions for implementing S.PSS and service design in developing countries to shape the future of product and service development and create sustainable economies. In Chapter 15, Dr Sophia N. Njeru, at Kirinyaga University in Kenya, describes her sustainable fashion safari (Kiswahili word for journey, expedition, or adventure) from 2000 to date in order to address the global fashion industry's challenge of highly unsustainable production and consumption systems. Dr Njeru's sustainable fashion safari, including her research, teaching, and practice, has opened many opportunities for acquiring and advancing her knowledge and skills about sustainability theories, models, research methods, pedagogy, publishing, grant writing, and collaborative research. In Chapter 16, Walter Chipambwa at Chinhoyi University of Technology in Zimbabwe talks about his journey in the textile industry since 2007, such as various projects in the industry (e.g., wet processing textile plant) and experiences in academia with broadened interests in upcycling, sustainability, and innovation in material science.

References

Fankhauser, S., Smith, S.M., Allen, M., Axelsson, K., Hale, T., Hepburn, C., Kendall, J.M., Khosla, R., Lezaun, J., Mitchell-Larson, E., & Obersteiner, M. (2022). The meaning of net zero and how to get it right. *Nature Climate Change*, 12(1), pp. 15–21.

HM Government (2021). Net Zero Strategy: Build Back Greener. Available at: https://www.gov.uk/government/publications/net-zero-strategy.

IPCC (2013). Climate Change 2013: The Physical Science Basis. Contribution of Working Group 1 to the Fifth Assessment Report of the Intergovernmental Panel on Climate Change. Cambridge University Press.

UN (2021). COP26: Together for our planet. Available at: https://www.un.org/en/climatechange/cop26.

UNFCCC (2015a). The Paris Agreement. Available at: https://www.un.org/en/clima techange/paris-agreement#:~:text=The%20Paris%20Agreement%20provides%20a, of%20the%20Sustainable%20Development%20Goals.

UNFCCC (2015b). Report on the structured expert dialogue on the 2013–2015 review. Note by the co-facilitators of the structured expert dialogue. Available on: https://digitallibrary.un.org/record/773092.

UNFCCC (2021). Glasgow Climate Pact. Available at: https://unfccc.int/sites/default/ files/resource/cma3_auv_2_cover%2520decision.pdf.

UNFCCC (2022). Sharm el-Sheikh Implementation Plan. Available at: https://unfccc. int/cop27.

PART I
SCIENCE AND ENGINEERING

1

MY ACADEMIC JOURNEY TO NET ZERO

From an environmental sociologist to the first co-chair of the 2050 Carbon Neutrality Commission

Sun-Jin Yun

Prior to majoring in energy and environmental policy

My undergraduate major was sociology. I was very concerned about society, social activities, public perception of social events, and social relationships between people and groups. Sociology seemed to be the right major to satisfy my concerns. When I was young, before I entered university, my dream was to be a teacher. Even though the Department of Sociology belonged to the College of Social Science, there was an opportunity to get a teacher's license by completing teaching subjects. I completed the course and became a social studies teacher at a girls' high school in Seoul in 1990. As a teacher, I created a special activity class called 'Current Affairs in the Environment.'

For the first time at that school, students were asked to present and discuss environmental news articles they found. There were two main reasons why I did this activity; these also led me to major in environmental and energy policy in graduate school.

One reason was the 1986 Chernobyl nuclear accident, which occurred when I was a junior in college. I was shocked and concerned by it, especially because my hometown was close to the nuclear power plants (NPPs) in Wolsung (a small city in Korea). Before Chernobyl, most Koreans saw nuclear power technology as a symbol of scientific progress and power for economic growth and advanced technology owned by developed countries. The risks associated with nuclear power technology were not recognized by the public, not even by people living near NPPs. Most local residents welcomed NPPs in their community because they thought they were modern electricity-generating factories that would bring economic growth to them and the country. I had not been educated about the dark side of NPPs in schools. The Chernobyl nuclear accident woke me up to it.

DOI: 10.4324/9781003380566-3

The other reason was a big social event in Korean education in 1989, a year before I became a teacher. The Korean Teachers Association was established in 1987. It promoted the creation of a teachers' union and advocated for truthful education despite the government's disapproval. The Korean Teachers Union (KTU), composed of public and private teachers, was established in 1989, and the government dismissed 1,490 union members. Even though the KTU was an illegal organization, it actively promoted various projects to realize truthful education, including environmental education projects that the government was not concerned about back then.

As soon as I became a teacher, I joined KTU and was baptized into environmental education. In Korea, most people were unaware of environmental problems before the 1988 Seoul Olympics. The government launched a massive environmental protection campaign because it did not want to show foreign guests Korea's polluted and damaged environment. The government was only interested in relocating factories in Seoul to provincial areas or making visible environmental problems invisible through environmental improvement projects. They were not interested in a more serious transformation of the structure that caused environmental problems. There was a slight improvement in environmental quality in Seoul, where the Olympics were held, but environmental problems continued to become more serious in Korea. KTU saw the importance of raising environmental awareness and conducted environmental education so teachers could spread it to students. KTU opened my eyes to environmental problems.

Climate change becomes a central focus of my research

Due to my concern regarding environmental and energy issues, I left my teaching career and went to graduate school in the USA to study the field further. Teaching is one of the most favoured jobs in Korea, especially for women. Even though my acquaintances discouraged me from quitting a stable job to start graduate studies with an uncertain future, I took the plunge and ventured into a new world. I was excited and grateful to learn something new every day in graduate school. It opened my eyes to a world I would never have known and allowed me to see it from a new perspective. I learned new concepts, including the energy system or energy path coined by Amory Lovins (1976), energy transition, energy poverty, energy welfare, environmental inequity and justice, sustainable development, various environmental discourses, and commons or common-pool resources. The plight of the poor in developing countries, suffering from both poverty and environmental problems, broke my heart. I specialized in the political economy of the environment and energy to understand the root causes of environmental and energy problems and to provide alternatives for an equitable and sustainable society.

Among the many environmental and energy issues, climate change has always interested me. Climate change is one of the most severe global environmental problems, and the burning of fossil fuels is a key contributor. This overlaps the issue at the intersection of environment and energy. In 1995, when I began my graduate studies, climate change did not have the importance it does now. Few students were pursuing climate change in their master's or doctoral thesis. Even environmentalists did not view climate change as a severe issue in South Korea. I was fortunate to have the opportunity to study at the Center for Energy and Environmental Policy at the University of Delaware, where climate change was considered very important. My master's thesis was titled 'An Alternative Policy Regime to Address Global Warming and Its Impacts on South Korea: Principles and Application', and my doctoral dissertation was titled 'Restructuring Political Economy in an Era of Global Energy and Environmental Change: Toward a Civil Society Approach to Promote a Climate-Sustainable Future.' I focused on equity and sustainability, specializing in civic engagement to address climate change. In a democratic society, people have a right to participate, demand and share information, voice their opinions, deliberate, and reach consensus. They are the origin of national power and decision making.

I was troubled by the fact that the individuals, groups, and countries that are more responsible for the climate crisis suffer less damage because they have the capital and technology to adapt. In contrast, individuals, groups, and countries that are less responsible suffer more severe impacts and damage. I was also troubled by the fact that the ongoing international climate change negotiations do not faithfully reflect the polluter pays principle or their historical responsibility, nor do they impose reasonable costs on the more responsible countries. Climate change is a problem caused by human socioeconomic activities, so humanity as a whole – especially individuals, groups, and countries that emit more greenhouse gases (GHGs) – have a responsibility to pay more and should take the lead in finding solutions. Climate justice has become an essential and integral concept in my studies and research.

Climate change requires a transition from a centralized energy system based on fossil fuels and nuclear to a decentralized one based on energy efficiency and renewable energy. The energy sector's reliance on fossil fuels is a major contributor to GHG emissions, so this energy transition is a necessary but insufficient condition for addressing climate change. In addition to my concerns about nuclear power technology, climate change has led me to delve into deeper issues related to the energy system. A sustainable society must consider the economy, environment, energy, and equity. Nuclear power is inherently risky and does not meet these concerns. In some countries, including South Korea, there is little disclosure and sharing of information about NPPs, leading some people to believe it is cheap energy. Proponents have promoted nuclear power as a carbon-free source of electricity generation and tried to expand it by using climate change as leverage. They ignore the adverse effects of extreme weather events caused by climate change on NPP operations (as shown in Ahmad, 2021). As I deepened my research, I argued that energy transition and climate change response should exclude nuclear power and fossil fuels.

Getting involved in the environmental movement as a professional

Upon returning from the USA with my PhD, I was fortunate enough to obtain a position as a professor in the Department of Public Administration at the University of Seoul in 2001. At that time, there were few climate experts in universities and civil society. I wrote about climate inequality and energy transition, topics that are rarely addressed in Korea. I consider the atmosphere one of the natural commons shared and used by all living organisms, including human beings. Thus, I wanted to know the traditional use and management of the commons to apply traditional wisdom and social institutions to the case of the atmospheric commons, albeit on a different scale.

I joined environmental organizations and became involved with environmental movements immediately after becoming a professor. My foreign colleagues, especially those majoring in environmental sociology, sometimes ask me why Korean experts, including professors, are actively involved in the environmental movement. This question may require scientific investigation to answer, but in my opinion, this is probably because lots of environmental issues require specialized scientific knowledge, and Korea has a cultural tradition in which intellectuals actively promote social opinions. As for myself, my undergraduate sociology major is deeply concerned with social activities and movements. Intellectuals must go beyond describing and explaining social phenomena and problems; they should engage in social movements to transform social institutions and policies based on expertise.

The first organization I joined was the Center for Energy Alternatives (CEA) of the Korea Federation for Environmental Movements (KFEM). The KFEM is the most significant environmental organization in Korea. It was founded in 1993 and has led many environmental movements, including the anti-nuclear movement (Yun & Dunlap, 2017). The CEA was a pioneering organization that pursued the energy transition. The first commercial nuclear operation in South Korea was the Kori Unit 1 in 1978. South Korea became the world's number one country in NPP density (installed capacity divided by land area) in 2005. The CEA was established in 2000 as an affiliated body of the KFEM, but separated from the KFEM in 2003, and changed its name to Energy Transition in 2005. In 2002, shortly after the introduction of the feed-in tariffs, it built the first citizen-owned solar power plant. Through the construction and operation of citizens' solar power plants, the CEA discovered institutional deficiencies and proposed policy alternatives to the government. As a result, it contributed to the realignment of renewable energy policies.

In 2008, the Lee Myung-bak government tried to promote the Pan-Korea Grand Waterway Project (PKGWP), which as a presidential candidate he had promised to implement. The crucial part of the PKGWP was the construction of a canal connecting the two largest rivers in Korea, the Han and the Nakdong, which are separated by mountains. However, the project faced social resistance, notably from 2,544 professors who launched the Professors' Organization for Movement Against Grand Korea Canal (POMAC). I served as an executive committee member of POMAC. This was the first time professors had organized themselves into a social voice on such a scale in Korea since the 1987 democratization movement. The Lee administration renamed it the Four Rivers *Restoration* Project (FRMP) and pushed ahead. Despite its name, this civil engineering project would build dams on the primary channels of four rivers, dredge the bottoms of these rivers to make them navigable, and convert the area around them into recreational spaces (Yun, 2014). The plan to connect two major rivers was dropped, but not much else changed.

The Lee government cited climate change as one of the main reasons for the FRMP, arguing that as climate change intensifies, the likelihood of floods and droughts will increase, so dams and dredging will increase 'water storage.' Without any scientific assessment of where, who, and how much climate change would affect them and how vulnerable they would be, 16 dams were built and dredged to a depth of six meters. I wrote a paper pointing out the undemocratic decision-making process and the resulting destruction of nature, as well as a second paper on the social responsibility of professionals in the process of promoting large-scale national development projects (Yun, 2014). These four rivers are the commons of all Korean people, including current and future generations. However, they were destroyed without sufficient social discussion, scientific evidence, or democratic decision-making processes. Despite the spread of algal blooms and the destruction of aquatic ecosystems on the Four Rivers, there is no political consensus for dam removal.

The role of the media is crucial, especially regarding energy and environmental issues like climate change. These issues require scientific knowledge and understanding. The deficit model assumes the general public will act rationally when provided with scientific knowledge. This is not entirely correct, but some public understanding of the science behind emerging phenomena, including climate change, remains essential. Knowledge produced by scientists is introduced and communicated to the general public through the media. What is reported in the media affects the public's understanding of science and behaviour. Therefore, I have been very interested in media coverage and have conducted various media analyses. After observing that the Korean media did not correctly report the problems with the FMRP, I conducted a content analysis of the 9pm news reports on KBS and MBC, which had the most influence on public opinion in Korea (Yun & Lee, 2014). The 9pm reports on the two broadcasters reduced the FMRP to a conflict between the two political parties rather than reporting its essential problems or core issues. No attention was paid to whether the FRMP would effectively combat climate change.

I was also interested in media coverage of climate change. Through the Global Research Network, an international collaborative research project supported by the National Research Foundation of Korea, I joined researchers from more than 20 countries in conducting a comparative analysis of how conservative, liberal, and economic newspapers cover climate change and what frames they use. Three key questions were asked: (i) How do the three selected representative newspapers describe climate change? (ii) What sources are they citing, and how do these sources vary across newspapers? (iii) What solutions are the newspapers proposing? The South Korean media reported climate change issues in a prognostic frame, focusing primarily on policy decisions and mitigation policies (Yun et al., 2012). Regardless of their ideological leanings, the government was the most cited source for all three Korean newspapers. However, there were differences in what each newspaper emphasized. The conservative and economic newspapers supported policies to maintain capitalist economic growth while preserving existing economic and social structures. The liberal newspaper called for aggressive GHG emissions reductions and was critical of high-risk, large-scale technologies such as nuclear power.

Participants in this global collaboration created an academic community called the COMparative POlicy Network (COMPON). Through media analysis, the collaborators identified key climate actors frequently mentioned in each country. They then surveyed these key climate policy actors and conducted a policy network analysis to see how they interacted by exchanging resources and information and which actors collaborated and conflicted. In the case of South Korea, government agencies were found to be the key actors mediating between business and the private sector. However, the government was not a monolithic entity but a collection of organizations with conflicting interests. Two distinct coalition networks, pro-growth, and pro-environment, existed in South Korea, with critical organizations leading their formation and maintenance. My study found that providing economic opportunities to actors in the environmental network and opening up the policymaking process can lead to more proactive climate action (Yun et al., 2014). So far, comparative studies have been conducted among COMPON participants (Kammerer et al., 2021; Karimo et al., 2022).

Towards a post-nuclear society after the 2011 Fukushima nuclear disaster

On the 11th of March 2011, one of the most serious nuclear accidents in human history occurred at the Fukushima NPP in Japan. Most South Koreans were shocked to learn that a nuclear accident had occurred in Japan, which was highly regarded for its nuclear safety management. The Fukushima disaster shattered the safety myth that NPPs never have accidents. However, despite their awareness of the risks associated with NPPs, many Koreans tended to accept the operation of NPPs as an inevitable, necessary evil.

During the Fukushima disaster, South Korea had 21 nuclear reactors in operation (Yun, 2015). In addition, seven reactors were under construction (Kori 2, 3, and 4; Shin-hanul 1 and 2; and Shin-wolsung 1 and 2), four reactors were in preparation for construction (Kori 5 and 6 and Hanul 3 and 4), and two reactors (Kori 7 and 8) had completed construction plans. In 2010 before the Fukushima nuclear disaster, nuclear power accounted for 12.2% of primary energy, 24.8% of total installed capacity, and 31.4% of total electricity generation. Even then, South Korea's NPP density ranked first in the world, as did the number of reactors, installed capacity, and people living in the areas surrounding NPPs. After constructing Kori 5 and 6, the nuclear risk would be concentrated around ten reactors, with more than 3.8 million people living within 30 kilometers of the plant. Even after the Fukushima disaster, the Lee government maintained its policy of expanding NPPs; however, the plan to locate NPPs in Samcheok, which had a confirmed construction site, was delayed because of local opposition. The plan to locate four reactors without a confirmed site was postponed. Instead, the Lee government approved the construction of ten coal-fired power plants (8.7 GW) on its last day in office in the name of stable electricity supply, despite touting 'low-carbon green growth' as the new national development strategy. Coal-fired power plants continue to be a source of social conflict because they generate fine dust, destroy landscapes, and make it challenging to combat climate change.

In the wake of the Fukushima nuclear disaster and the irregularities in the nuclear industry revealed by the Park Geun-hye administration, including the use of non-genuine parts, all but one of the major presidential candidates in the 2017 presidential election made nuclear phase-out a campaign promise. Moon Jae-in won the election on a nuclear phase-out pledge, which received the highest support in the online policy mall that ran during the election. As a candidate, Moon promised to suspend the construction of all nuclear power plants under construction. However, on the 19th of June 2017 at a proclamation ceremony for the permanent suspension of Wolsong 1, he suggested that the suspension of Shin-Kori 5 and 6, which were then under construction, be decided through public discussion. He formed a public deliberation committee consisting of nine experts, which organized a participatory deliberation group of 500 ordinary citizens. After three months of the public discussion process, the final survey of the participatory deliberation group showed 59.5% in favor of resuming construction.

After the Fukushima nuclear disaster, I participated in the founding of the Korean Professors' Organization for a Post Nuclear energy Society (K-POPONS) and served as its executive chair. Since then, I have been more actively involved in public lectures, publishing papers in domestic and international journals and books, and attending and presenting at numerous debates and seminars about nuclear phase-out. I also participated in the launch of Seoul's One Less Nuclear Power Plant (OLNPP) comprehensive plan in 2012, led by Mayor Park Won-soon. From 2017 to 2021, I chaired the Executive Committee for OLNPP, which was renamed as the Seoul Energy Policy Committee in 2019. Seoul has tried many social experiments to reduce its dependence on nuclear power and fossil fuels and has developed and implemented various programs (Ahn, 2017; Byrne & Yun, 2017; Yun, 2017; Gunderson & Yun, 2021). I have also delivered lessons from OLNPP to other societies, especially Taiwan, through several seminars and public lectures.

I participated in the public consultation process as a member of the expert panel, submitting expertise on why the construction of Shin-Kori 5–6 should be stopped and participating in a two-day debate. The public consultation on Shin-Kori 5–6 was the first attempt to realize deliberative democracy based on citizen participation in Korean society, especially nuclear policy. It can be evaluated as a successful social experiment. It confirmed the possibility of public consultation to resolve social conflicts by providing relevant information to ordinary citizens who had been excluded from the policy-making process for lack of expertise. However, nuclear proponents, who already had strong organizational power and resources, mobilized their resources to defend their interests. The space of public consultation was also used to their advantage. The anti-nuclear camp was short on money and time and operated in loose coalitions based on voluntarism, resulting in poor information production, and sharing. This was one reason why citizen participants who agreed to resume construction were concerned about the costs invested so far but did not consider that future costs would be much higher. The Moon government accepted the suggestions of the public deliberation committee. It announced the Energy Transition and Nuclear Phase-out Roadmap in October 2017, cancelled plans to build new NPPs and did not extend the life of old NPPs.

Members of the expert panel on the suspension of reactor construction, including myself, agreed that an organization bringing together experts, businesses, and citizens was needed. This group could go beyond criticizing the problems of NPPs and press the government to move forward with the energy transition systematically, suggest policy and institutional alternatives, and provide useful information to ordinary citizens to win public support. In February 2018, Energy Transition Forum Korea was launched, and I became a registered board member. Unlike other civil society organizations, it included businesses because we believed that companies interested in the energy transition and working in the fields of renewable energy and energy efficiency could be the driving force behind the new transition. However, Korea, which has the lowest share of electricity from renewable energy among OECD (Organisation for Economic Cooperation and Development) countries, faces a wide range of obstacles to the energy transition, including low electricity prices, the Korea

Electric Power Corporation's monopolistic electricity market, and overreliance on nuclear power and fossil fuels. Despite the efforts of Energy Transition Forum Korea, the energy transition has been slow.

Serving as co-chair of the 2050 Carbon Neutrality Commission

In 2018, the IPCC (Intergovernmental Panel on Climate Change) released its 'Special Report on Global Warming 1.5°C,' recommending net zero by 2050 and a 45% reduction in CO_2 emissions by 2030 from 2010 levels in order to meet the 1.5°C goal instead of 2°C. Following the Climate Action Summit held at the UN (United Nations) in September 2019, countries worldwide declared their commitment to net zero. On 28 October 2020, Moon declared carbon neutrality by 2050 in his speech at the National Assembly; on the 10th of December, he held a national vision declaration ceremony with the slogan 'Carbon neutrality by 2050 before it is too late.' Achieving carbon neutrality in 2050 required an upward revision of each country's 2030 nationally determined contribution (NDC). These were submitted before and after adopting the Paris Agreement, which did not have a net zero goal. Following the declaration of net zero, many countries strengthened their 2030 NDCs. The Moon government submitted a 2050 long-term, low-carbon GHG emission development strategy to the United Nations Framework Convention on Climate Change on the 30th of December 2020. Moon pledged to revise the 2030 NDC upward within his term. On the 22nd of April 2021, US President Joe Biden hosted the World Climate Summit and called on countries to strengthen their 2030 NDCs. He announced a USA proposal to reduce GHG emissions by 50–52% below 2005 levels. At the summit, Moon pledged to raise Korea's 2030 NDC by the end of 2021. However, this timeline was pushed back to the Conference of Parties 21 in early November in a joint declaration during Moon's visit to the USA in May 2021.

On the 29th of May 2021, the 2050 Carbon Neutrality Commission (CNC), a joint public-private governance body, was launched. The CNC is the first committee in the world to include the term 'carbon neutrality.' The CNC was to be composed of 50 to 100 members, including the prime minister and a civilian co-chair. The heads of 18 central administrative agencies are ex officio members. In contrast the civilian members appointed by the president include experts in climate, energy, economy, industry, and technology, as well as people from various walks of life, including civil society organizations, youth organizations, businesses, and labor unions. The first CNC had 77 members.

When I was offered the position of co-chair, I was hesitant. It was evident that both environmental groups and industry would criticize me. In a governance body composed of diverse stakeholders, no one side's opinion can be accepted unilaterally. Domestically, environmental groups would want a faster and more aggressive response, while industries based on high energy consumption would demand a slower response. As co-chair, I led the national decision-making process of establishing the 2050 carbon neutrality scenario (CNS). I was constantly thinking about how to reach a consensus in the face of opposing sides that would strengthen the 2030 NDC. The co-chair is an unpaid adjunct position, which meant I had to juggle my chair duties while maintaining my teaching position. As a scholar who has studied the best practices for governance and as someone who has participated in several governance organizations, I was determined to do my best to get the CNC right.

The CNC was given fewer than five months to deliberate and decide on a CNS and the enhanced 2030 NDC. Within that time, the CNC had to gather and synthesize a wide range of opinions from inside and outside the committee before making a final decision. The CNC amended early CNSs and the enhanced 2030 NDC plan submitted by the government through subcommittees, general planning committees, expert committee meetings, a scenario task force, and an NDC task force. Before it disclosed the three draft CNSs on 5 August 2021, there had been 54 intensive internal meetings. Initially, two CNSs were submitted by the government, with coal and gas in the first scenario and gas in the second scenario in the energy transformation sector, with heavy use of carbon capture, utilization, and storage (CCUS). However, after internal discussions in the CNC, a third scenario with no fossil fuels for electricity generation was added to the draft.

The inclusion of coal-fired power in the first scenario was met with strong opposition from environmental groups. Scenarios were meant to show how different outcomes emerge under different assumptions. While the inclusion of coal-fired power generation demonstrated that carbon neutrality would be challenging to achieve domestically and that CCUS would need to be applied at a significant scale, the fact that it was included was of interest only to them. The scenarios presented were drafts and were intended to solicit feedback. After public feedback, the first scenario was removed, and the other two scenarios were strengthened to achieve carbon neutrality domestically. Revising the scenario included public discussion through the Citizens' Council for 2050 Carbon Neutrality (CCCN), meetings with stakeholder councils, and forums for the general public.

The CNC operated the CCCN, composed of members of the public and designed to promote discussion about a decarbonized society. The CCCN comprised more than 500 people aged 15 and older nationwide, with quota sampling by gender, age, and region. Generally, when public opinion polls are conducted or deliberative public participation bodies are organized in Korea, people over age 19 are targeted. However, the CCCN lowered the age to 15 to hear the voices of youth – the future generation. The CNC also needed to gather opinions from outside the commission. The CNC established consultative meetings with industry, labor and farmers, civil society organizations, youth organizations, and local governments to hear and collect opinions from direct stakeholder groups. More than 20 in-person meetings were held with 115 associations and organizations, and 94 organizations submitted written positions to the CNC. The CNC also held two online and offline meetings for the general public, then finalized the CNS and upgraded the 2030 NDC.

Continuing research and education for a net zero society

The 2050 CNS and 2030 NDC are neither sufficient nor complete; therefore, they must be continually supplemented. With the change of government, I stepped down as co-chair, and now, as a professor, I intend to dedicate myself to 2050 carbon neutrality through education, research, and social activism. A 2050 carbon-neutral society is not possible without a massive transition from the current carbon civilization. All laws, policies, systems, and people's lifestyles and mindsets must change. There is a lot to research, and I realize I now want to study more carefully how people's daily practices should change. I would like to find alternatives through field research on how to increase social acceptance of renewable energy.

The first Basic Plan for Carbon Neutrality and Green Growth, established by the current Yoon Seok-yeol administration, faced strong opposition from civil society for a variety of reasons: it lacked a proper public consultation process, a long-term plan beyond the plan's target year of 2042, and a financing plan to realize carbon neutrality. Despite the urgent need to reduce GHGs, the Yoon government made only a quarter of the reductions, leaving most of them to the next government. It lowered the industrial sector reduction target from 14.5% to 11.4%, increased the share of nuclear power, and lowered the share of renewable energy. The Yoon government has abandoned the Moon government's nuclear phase-out policy and is promoting NPP expansion in the name of low-carbon emissions. Transitioning to a sustainable, equitable, and just carbon-neutral society requires more than a technological transition. It requires a sociotechnical transformation and more specific strategies for achieving it. I intend to continue to dedicate my work to charting such a transition path.

References

Ahmad, A. (2021). Increase in frequency of nuclear power outages due to changing climate, *Nature Energy*, 6: 755–762. https://doi.org/10.1038/s41560-021-00849-y.

Ahn, B. (2017). One Less Nuclear Power Plant: A Case Study of Seoul Megacity, in *Reframing Urban Energy Policy*, Seoul Metropolitan Government.

Byrne, J., & Yun, S. (2017). Achieving a Democratic and Sustainable Energy Future: Energy Justice and Community Renewable Energy Tools at Work in the OLNPP Strategy, in *Reframing Urban Energy Policy*, Seoul Metropolitan Government.

Gunderson, R., & Yun, S. (2021). Building energy democracy to mend ecological and epistemic rifts: An environmental sociological examination of Seoul's One Less

Nuclear Power Plant initiative, *Policy Studies Journal*, 72: 36–44. https://doi.org/10.1016/j.erss.2020.101884.

Lovins, A. (1976). The road not taken, *Foreign Affairs*, 55: 65–96.

Kammerer, M., Wagner, P., Gronow, A., Ylä-Anttila, T., Fisher, D., & Yun, S. (2021). What explains collaboration in high and low conflict contexts? Comparing climate change policy networks in four countries, *Policy Studies Journal*, 49 (4): 1065–1086. https://doi.org/10.1111/psj.12422.

Karimo, A. *et al.* (2022). Shared positions on divisive beliefs explain interorganizational collaboration: Evidence from climate change policy subsystems in 11 countries, *Journal of Public Administration Research and Theory*, 1–13. https://doi.org/10.1093/jopart/muac031.

Yun, S. (2014). Experts' social responsibility in the process of large-scale nature-transforming national projects: Focusing on the case of the four major rivers restoration project in Korea, *Journal of Asian Sociology*, 43 (1), 109–141. doi:10.21588/dns.2014.43.1.005.

Yun, S. (2015). *Korea's Nuclear Policy – Past, Present, Future, KAS Journal on Contemporary Korean Affairs: Environmental Policy in South Korea – Problems and Perspectives*. Konrad-Adenauer-Stiftung Korea Office.

Yun, S. (2017). Citizen participation-based energy transition experiments in a megacity: The case of the One Less Nuclear Power Plants in Seoul, South Korea, in Chou, K. (ed.), *Energy Transition in East Asia: A Social Science Perspective*. Routledge Taylor & Francis Group.

Yun, S., & Dunlap, E.R. (2017). *Environmental Movements in Korea: A Sourcebook*. The Academy of Korean Studies.

Yun, S., Ku, D., & Han, J. (2014). Climate policy networks in South Korea: Alliances and conflicts, *Climate Policy*, 14 (2): 283–301. doi:10.1080/14693062.2013.831240.

Yun, S., Ku, D., Park, N., & Han, J. (2012). A comparative analysis of South Korean newspaper coverage on climate change: Focusing on conservative, progressive, and economic newspapers, *Development & Society*, 41 (2): 201–228. https://hdl.handle.net/10371/86765.

Yun, S., & Lee, D. (2014). Agenda setting and frame of TV news about 4 major rivers project in Korea, *ECO*, 14 (1): 7–62 (in Korean).

2

THE ROLE OF DISCIPLINARY, INTERDISCIPLINARY, AND TRANSDISCIPLINARY EDUCATION IN SUSTAINABILITY SCIENCE

Experiences from my professional journey

Jagdeep Singh

Early education in electrical engineering

I was born and raised in India, where I completed my schooling at my village school in the foothills of the Himalayas. In a country with 1.3 billion people and millions of students competing to get a seat to study in India's elite educational institutions, deciding on what to study and where to study are serious questions during the last two years of school. This is because preparations for the future profession start about two years before school ends (meaning coaching classes or tuition to prepare for qualification entrance examinations). So, most students know what they want to be in the future. However, in my village school in the early 2000s, with a lack of information about this issue, life was relatively simple. I did not know clearly what to do next. By default, I planned to be admitted into the Bachelor of Science programme at the local university college, and complete my education to become a high school teacher.

While this was the plan, I also took an engineering entrance exam. Based on my exam results, I was asked to select an engineering branch for my Bachelor of Technology education. My parents did not know much about engineering, so I was asked to make this decision at the age of 17 years. In the early 2000s, there was a boom in information technology (IT) jobs in India (and indeed worldwide). So, anticipating excellent job prospects after education, some people suggested admission into the IT engineering programme. However, I had no interest in studying IT. Since I enjoyed studying physics, I selected electrical engineering and enrolled in the four-year Bachelor of Technology course at an engineering college.

That was how in 2002 my journey as an engineer started. I enjoyed studying the subject, and it was full of new knowledge, ranging from simple topics, such as Ohm's law, which dictates how the current flow in an electric circuit is affected

DOI: 10.4324/9781003380566-4

by its resistance and the applied voltage, to more complex concepts, such as the working phenomena behind a transistor, induction motor, transformer, or modern electric power transmission systems. Most of the study programme during the first three years was focused on mathematical concepts with little demonstration of these concepts in practice (mainly in the practical workshops).

It was challenging to learn about so many 'invisible' concepts such as electric flux, electric current, electromagnetic field, electric potential and so on. I remember one instance when, in one of the first-year workshops, I asked my teacher how the current/power transfer takes place from one transformer coil to another since they are not physically connected. That day, I understood the concept of electric and magnetic flux. Also, I was finally able to fully understand what was taught during my high school years about the behaviour of an electron in an electric field. I would later realise the usefulness of these mathematical and analytical concepts in my interdisciplinary journey in the field of sustainability science.

Work as an engineer trainee for six months

After completing my engineering education, I joined a manufacturing plant as a maintenance engineer. There, I experienced the real-world applications of electrical and mechanical engineering in various plant utilities such as heating, ventilation, and air-conditioning (HVAC), emergency power supplies, and water purification units. For the first time, I saw how electric and mechanical transducers are employed to sense temperature, pressure, or flow to control and operate valves and how programmable logic controllers are programmed to perform different functions. During this job, after practically working with electrical and mechanical machines, the concepts I learned during my engineering education were way too advanced and appeared almost useless in daily routine work. Since I wanted to study further, I quit this job and started preparing for the entrance exam for my master's studies in India. Over the years, however, this work experience has helped me in many ways to reflect critically on the life cycle inventory data on energy and material flows collected from companies as part of the life cycle assessment (LCA) studies during my doctoral and postdoctoral research.

Those three months of work in rural India

The master's entrance exam was held in February, with the results announced in March. My score on the entrance exam was not satisfactory. I was unsure about getting a position in the power electronics master's programme (the subject I wanted to pursue further) at my desired university – the Indian Institute of Technology (IIT). Therefore, based on my score, I applied for a Master of Technology in energy studies at IIT Delhi. I also applied for other courses at several other universities across India.

Meanwhile, whilst I was waiting for responses to my applications, I took a job as a Project Engineer in a governmental consulting firm. My job was to visit the sites of the newly installed electric power distribution network and inspect the work. During these three months of employment, I visited several rural villages in the State of Rajasthan. These villages were among the last locations in India to get access to electricity (as of the year 2007). Although I was born and raised in a small village, the village was connected to the electricity grid around 1978.

Visiting these remote villages was a culture shock, with acute poverty, illiteracy, and no access to clean water and electricity. The villagers told us about the unethical work practices of the contractors who laid the transmission network and how they were paid little or not paid at all for their work installing the electric poles and wires. This was difficult for me to hear. I stayed in a city about 80 km from these remote villages during this job. Although this city was considered an education hub of India, these villages appeared to be separated by a thousand miles from a socio-economic viewpoint. Travelling daily to these villages by road transport and sometimes walking and witnessing their situation was emotionally exhausting and physically tiring. I never realised how this experience would profoundly impact my personality and my educational and professional journey in the coming years.

Master of Technology in Energy Studies

Luckily, I was shortlisted for an interview at IIT Delhi. My name on the list of interviewees was second to last for the Master of Technology in Energy Studies programme. I was asked some tricky questions related to electric circuits at the interview. I was confident that my answer was correct. When the results were announced the following day, I was accepted into the Energy Studies programme. The programme focused on renewable energy technologies and their role in solving societal energy challenges. This interdisciplinary programme comprised students with multiple disciplinary backgrounds, such as physics, chemistry, and engineering. The programme had a set of interesting courses from different disciplines, such as Solar Architecture, Fuel Technology, Energy, Ecology and Environment, Energy Conversion, Energy Conservation, Heat Transfer, Non-conventional Sources of Energy, Economics and Planning of Energy Systems, Quantitative Methods for Energy Management and Planning, and Power Generation, Transition and Distribution. There were lots of practical workshops where we were taught how to find out the calorific value of different fuels, how to determine the knocking numbers of petroleum and biodiesel, how to analyse different particulate matters in the fumes of different types of fuels, how to measure the radiation levels, and the efficiency of a photovoltaic module or solar heater, and so on.

The lectures that shaped my professional journey

Some of these courses were technological, and others focused on socio-economic and ecological aspects. As an engineer, I was very interested in understanding these aspects. Some lectures on the sustainability of energy (both non-renewable and renewable) impacted me deeply. One such lecture was on how the Kyoto Protocol describes the 2008 Intergovernmental Panel on Climate Change's (IPCC) planned mechanism to reduce carbon emissions through carbon credit schemes in developing countries. An example was drawn from the Montreal Protocol – a successful global-level effort to reduce chlorofluorocarbons (CFCs) in the atmosphere damaging the ozone layer. This lecture presented the current state of global emissions at that time and the carbon emission reductions needed in future to achieve climate targets. The number presented in the lecture was quite depressing. On the one hand, emissions were expected to increase because consumption in the developed nations was (and still is) on the rise. On the other hand, a significant global population in developing nations does not have basic amenities for sustenance. Moreover, there was no consensus globally to reduce carbon emissions.

In another course, I learned how global electricity production is dependent on coal or oil, how the current renewable technologies (such as photovoltaic and wind power) are competing with the incumbent power systems that are heavily subsidised, and how the income gap in the high-, middle- and low-income classes are increasing, making matters worse from a sustainability perspective.

These lectures reminded me of my previous experience working in the rural parts of India. I also revisited some of the subjects I learned during my engineering education. For example, the overall energy efficiency of a thermal power plant is less than 30%. Most of the energy/exergy losses take place in: (i) the boiler, where energy conversion takes place from heat to steam, (ii) the turbine, where energy from steam is converted to mechanical (rotational) energy, and (iii) the electric transmission over long distances. In one of the lectures on heat transfer, it was clear that a 1% increase in the boiler's efficiency in a thermal power plant can massively reduce carbon emissions. All these numbers made more sense to me beyond their technical contexts (i.e., the social and environmental). I was clear about what to pursue in my thesis project and career afterwards.

As my thesis project, I decided to work on passive building design in different Indian climatic conditions – buildings that can self-regulate their inside temperature to a human comfort level of 15 to 25 degrees Celsius without using an external power source. It involved design aspects of a building's walls, roof, floor insulation, and windows. I worked with solar radiation data from selected climates in Indian cities to understand what building design features can support passive heating or cooling. I used the TRNSYS®

software package to simulate different building designs in the selected climatic conditions, and the archived solar radiation data included in the programme for all the climatic conditions tested. I measured the hourly solar radiation and ambient temperature for New Delhi (at the IIT Delhi campus) for two months to understand how the sun's location affects the thermal radiation and temperature levels. I learned a lot during this project since the project involved understanding heat transfer through different media under varied internal and external temperatures and ventilation conditions. I pursued an out-of-programme course called 'Computer Programming and Its Application Using C Language' to understand computer programming. I found this course very difficult and uninteresting, and I barely passed it with a low grade. However, I repeatedly found it to be helpful in the ensuing years and was thankful for having pursued it.

While working on my thesis, I could appreciate my old home during my childhood, which was made using local materials and designed as per the local hot climatic conditions. Over the years, most of the old constructions in my village have been replaced by so-called 'modern' concrete and brick-made homes. These homes warm up faster, store heat in their structure, and irradiate the stored heat after sunset. This is one of the apparent reasons for the rising electricity demand to ventilate or air-condition homes in hot climates in India. My thesis work inspired me to build a passive home in my village using local raw materials and without any external power source to maintain its inside temperature to a human comfort level (although I am yet to realise this goal).

My job as university teacher for ten months

After completing my master's degree, I became a university teacher. I was suddenly back to teaching engineering subjects. I taught basic electronics, electromagnetic field theory, an in-depth course based on the Maxwell equations, and microprocessor 8085 architecture. I found the 'computer programming and its application' course very useful while teaching microprocessor architecture and its programming. I view this part of my career as a brief but essential pause in my professional journey. During this period, I realised I should pursue a career in sustainability science. So, I started applying for doctoral positions. I came across the Erasmus Mundus India4EU programme and applied for it. I chose the doctoral position offered at the Politecnico Di Torino, Italy and the KTH Royal Institute of Technology Stockholm, Sweden. Again, my hopes and expectations were low because there was only one open position for India within this programme. I was surprised when, after a few months, I received a call from Sweden informing me that my application had been selected and that if I was still interested in pursuing my doctoral studies at KTH, it would be forwarded to the EU. I happily accepted this offer. I resigned from my job as a teacher after ten months. My next destination in my professional journey was Stockholm, Sweden, where I stayed for six years.

Doctoral education in industrial ecology

I joined the Division of Industrial Ecology at KTH Royal Institute of Technology to pursue a Licentiate of Technology (Industrial Doctorate) and a Doctor of Philosophy (PhD) in Industrial Ecology. Industrial ecology is an interdisciplinary research field integrating various disciplinary perspectives from engineering, environmental science, and social science to contribute to sustainable development. It aims to transform industrial systems to resemble natural ecosystems to optimise the use of natural resources, energy, and capital. The research within the field of industrial ecology is, therefore, mainly problem-driven and is conducted at the interface of the technological, ecological, and socio-economic systems to address the sustainability challenges facing society. Some of the critical sustainability principles of the field of industrial ecology are based on: (i) closing and slowing the material loops, (ii) efficient energy utilisation, and (iii) sustainable product and industrial design focusing on industrial symbiosis, dematerialisation, material substitution, product maintenance, reuse, recovery, remanufacturing, and waste management. Industrial ecology employs systems thinking and life cycle thinking to analyse the stocks and flows of materials and energy in industrial and consumer activities and their impacts on the environment and society.

On my first day at KTH, I met my colleagues from different parts of the world – some who had already been there for a few years, and a few who had recently joined the same doctoral programme from Brazil, Bangladesh, and China. When I applied for the doctoral programme, I submitted a project proposal on the product design of electrical and electronic goods from an environmental perspective. However, I was presented with an updated research plan focusing on life cycle assessment and modelling of waste management systems. I started my doctoral studies by exploring the MATLAB Simulink software ORWARE (ORganic WAste REsearch). After analysing the software's programming structure and mathematical equations, I realised that I alone could not update it as per the latest waste management technologies (e.g., emission factors of incinerators, collection trucks, etc.). Consequently, I developed an updated research plan during the initial literature review and doctoral supervision meetings.

The new research plan aimed to address the sustainability challenges in product design, resource extraction, production, and consumption from a waste management perspective (an end-of-life perspective on the life cycle of products). The main objectives of the research within this broad, abstract theme evolved over the coming years. This is because the research aims and objectives were not fixed. It was a challenging task. However, I was not alone: I was part of a small research group, with two fellow doctoral students – Rafael Laurenti and Rajib Sinha – and our doctoral supervisor, Prof Björn Frostell. This research group focused on overarching sustainability challenges facing product design systems, production, consumption, and waste management. Rafael Laurenti's research focused on this system from a product design perspective (a start-of-life perspective on the life cycle of products), and Rajib Sinha's research focused on modelling and simulating products' life cycle in this system under different sustainability scenarios.

Over the years, we had a fantastic journey of learning from each other. Together, we collaborated on several research projects and wrote a number of journal manuscripts. Some of these projects included, for example, (i) sustainability challenges to current waste management system paradigm (Singh et al., 2014), (ii) addressing the unintended negative consequences of improvement actions in production and consumption systems (Laurenti, Singh, et al., 2016; Laurenti, Sinha, et al., 2016), (iii) pervasive challenges to sustainability by design of electronic products (Laurenti et al., 2015), (iv) challenges to the upcoming circular economy to recover resources from post-consumer waste (Singh & Ordoñez, 2016), and (v) identifying ways to close the metal flow loops in the global mobile phone product system (Sinha et al., 2016). The first journal article I led was entitled 'Progress and challenges to the global waste management system' (Singh et al., 2014). It was published in the *Waste Management & Research Journal*'s special volume on wastes and cities. This publication was selected as 'the editors' choice article of the month'. The article was available as Open Access and distributed worldwide to waste management professionals and researchers. One year later, the same article won the International Solid Waste Associate (ISWA) publication award at the ISWA World Congress in Antwerp, Belgium.

During my doctoral research, I defended two doctoral-level theses – a Licentiate thesis (Singh, 2014) and a Doctoral (Singh, 2016) thesis in 2013 and 2016, respectively. While waiting for my Doctoral defence seminar, I accepted a research call offer by the KTH Business Alliance. The project involved collaborating with an entrepreneur to design, implement and evaluate the circular economy (CE) based plastic bag waste management business model from the social, economic, and environmental perspectives (Singh & Cooper, 2017). This project was one of my more successful collaborations beyond my research group at KTH. The results of this study were showcased in the Swedish newspaper *Debatt* (The Debate). This was my first appearance in a newspaper debate article. I carried this research project to my next destination in my research journey at the Centre for Industrial Energy, Material and Products (CIE-MAP) at Nottingham Trent University (NTU), England.

Postdoctoral research: research fellow in sustainable business

During my one-year research at NTU, I met very inspiring colleagues, such as Prof Tim Cooper, Dr Christine Cole, Dr Alex Gnanapragasam and Dr Kyungeun Sung (the editor of this book). My research involved identifying and evaluating environmentally sustainable business practices focusing on resource efficiency and the circular economy. We analysed more than 550 market offerings for consumer products based on: (i) the duration of products' guarantees or warranty offered with sale or lease, (ii) the provision of spare parts or service of repair during the products' use phase, and (iii) provision for the end-of-life product or spare-part take back system (Singh, Cooper, et al., 2019). In this work, I devised an innovative quantitative method to evaluate these products' circularity index. My interdisciplinary educational background and expertise in quantitative methods assisted with this work. In another interdisciplinary research project (mentioned above), I employed a life cycle assessment to evaluate a circular business model's carbon, water, and energy footprints for plastic waste management in Sweden (Singh & Cooper, 2017).

I collaborated with Dr Christine Cole on a project on barriers to resource reuse and recovery of the UK waste electrical and electronic equipment (WEEE) sector (Cole et al., 2018, 2019b). We also analysed how the objectives of the WEEE directives in promoting the movement up the waste hierarchy can be achieved in the UK (Cole et al., 2019a). We also explored consumers' perspectives on products' longevity and reliability (Gnanapragasam et al., 2017, 2018). I continued to collaborate with Dr Rafael Laurenti and Dr Rajib Sinha on research on the circular economy. We analysed the social embeddedness of the circular economy (Laurenti et al., 2018) and proposed an element flow analysis approach to support the circular economy (Sinha et al., 2020). Kyungeun Sung (who was pursuing her doctoral studies at that time) introduced me to a new research horizon – resource upcycling – that I continued to explore and develop at the International Institute for Industrial Environmental Economics (IIIEE) at Lund University after finishing my postdoctoral research at NTU, England.

Second postdoc: postdoctoral researcher in sharing economy and upcycling and repair DIY (do-it-yourself) activities

At IIIEE, I evaluated the social, economic, and environmental impacts of the sharing economy or collaborative consumption in the accommodation, mobility, and physical goods sectors. The research involved systems analysis of new interactions and causal mechanisms within the sharing economy and their broad sustainability impacts on society. Together with the research team, I developed a framework to evaluate the sustainability impacts of the sharing economy. Together with my colleagues, I explored the following research projects: challenges and research need in evaluating these impacts using input-output analysis (Plepys & Singh, 2019), the sharing economy state of research (Laurenti et al., 2019), systematic framework to assess the social impacts of sharing platforms (Curtis et al., 2020), and the urban sharing in Malmö, Sweden (Palgan et al., 2019), Amsterdam, Netherlands (Mont et al., 2019), and Toronto, Canada (Mont et al., 2020).

Along with this research project, I analysed the upcycling activities at DIY bike repair cafes in Sweden, Spain, and Switzerland from a sustainability perspective (Singh et al., 2021). I collaborated with Kyungeun Sung on various research activities in upcycling research. These included co-organising the 1st International Upcycling Symposium, co-editing a special issue on upcycling research, and co-editing a book on upcycling research and practices worldwide. These collaborations led to the publication of several research articles, such as exploring challenges and opportunities for scaling up the wood and textile upcycling businesses in the UK (Singh et al., 2019), scaling up British fashion upcycling businesses (Sung et al., 2020), systems approach to scaling up global upcycling (Singh & Sung, 2021), state-of-the-art upcycling research and practice (Sung et al., 2021), and upcycling, *jugaad*, and repair cafes for prosumption (Singh & Arora, 2021).

Researcher in environmental science

After completing my second postdoc at IIIEE, I joined the Centre for Environmental and Climate Science (CEC) at Lund University. Here, my research is focused on investigating challenges and opportunities associated with deriving biomass from agricultural land in Sweden for bioenergy purposes and mitigating climate change. The work involved formulating future scenarios for agricultural bioenergy/biomass production together with key stakeholders in the bioenergy production and consumption system in Sweden. This was quite different from my previous research. However, my educational background in Energy Studies proved very useful in understanding the system of bioenergy production and consumption and the associated research tasks.

Furthermore, I approached various research objectives by employing various qualitative (e.g., semi-structured interviews, user surveys, stakeholder workshops, causal loop diagrams, content analysis, thematic analysis, etc.) and quantitative methods (such as system dynamics and life cycle assessment) that I had employed during my doctoral and postdoctoral research. Moreover, based on my personal life experiences in the village where I grew up, pursued my school education, and helped my family in farming activities, I could closely relate to the issues of biodiversity loss, impacts on soil quality, water scarcity, and land use change. This helped me formulate interview guides and workshops to explore backcasting scenarios for the Swedish bioenergy sector with the local farmers and other key stakeholders. In my current research activities, I am continuing to employ the research methods used in my previous research, such as causal loop diagrams, life cycle assessment, and thematic analysis, as well as methods that are new to me, such as backcasting scenarios and biodiversity and ecosystem services impacts framework.

Future research plans

In my current position as Associate Professor in Environmental Science at the Centre for Environmental and Climate Science (CEC), Lund University, I am advancing research on sustainable production and consumption in two research areas. Firstly, I plan to explore consumer-led DIY repair and upcycling activities and their sustainability potential under various policy conditions and business model scenarios. Secondly, I want to analyse the potential of urban farming to advance environmental sustainability, promote biodiversity in an urban context, and reduce individual consumption due to changed user behaviour among the citizens who spend a significant amount of time on urban farming activities. In both research areas, I will analyse the citizens' individual time use in conducting these activities and their impact on their daily consumption. In addition, I am working on agri-symbiosis analysis exploring the cascade utilisation of material and energy resources in agricultural and forest resources to generate an overall sustainable value chain from a national/regional perspective. I am also analysing unintended consequences of bioenergy production and consumption systems in a global sustainability context. This is to address the direct and indirect negative social, economic, and environmental impacts of bioenergy production and consumption. In these research projects, I will continue expanding my research expertise in qualitative and quantitative methods for data collection and analysis to address societal sustainability challenges.

Conclusions

This chapter presented an overview of my educational and ongoing professional journey towards an interdisciplinary research field of sustainability science. Some key moments and decisions that shaped my educational and professional journey have been highlighted. The chapter emphasised how my previous education and job experiences have immensely enriched this journey.

References

Cole, C., Gnanapragasam, A., Cooper, T., & Singh, J. (2019a). An assessment of achievements of the WEEE Directive in promoting movement up the waste hierarchy: experiences in the UK. *Waste Management*, 87, 417–427. https://doi.org/10.1016/J.WASMAN.2019.01.046.

Cole, C., Gnanapragasam, A., Cooper, T., & Singh, J. (2019b). Assessing barriers to reuse of electrical and electronic equipment, a UK perspective. *Resources, Conservation & Recycling*: X, 1, 100004. https://doi.org/10.1016/J.RCRX.2019.100004.

Cole, C., Gnanapragasam, A., Singh, J., & Cooper, T. (2018). Enhancing reuse and resource recovery of electrical and electronic equipment with reverse logistics to meet carbon reduction targets. *Procedia CIRP*, 69, 980–985. https://doi.org/10.1016/j.procir.2017.11.019.

Curtis, S. K., Singh, J., Mont, O., & Kessler, A. (2020). Systematic framework to assess social impacts of sharing platforms: Synthesising literature and stakeholder perspectives to arrive at a framework and practice-oriented tool. *PLoS ONE*, 15 (10), e0240373. https://doi.org/10.1371/journal.pone.0240373.

Gnanapragasam, A., Cole, C., Singh, J., & Cooper, T. (2018). Consumer perspectives on longevity and reliability: a national study of purchasing factors across eighteen product categories. *Procedia CIRP*, 69, 910–915. https://doi.org/10.1016/j.procir. 2017.11.151.

Gnanapragasam, A., Oguchi, M., Cole, C., & Cooper, T. (2017). Consumer expectations of product lifetimes around the world: a review of global research findings and methods. In C. Bakker & R. Mugge (Eds.), *Product Lifetimes and the Environment 2017*. IOS Press. https://doi.org/10.3233/978-1-61499-820-4-464.

Laurenti, R., Singh, J., Cotrim, J. M., Toni, M., & Sinha, R. (2019). Characterizing the sharing economy state of the research: A systematic map. *Sustainability*, 11 (20). https://doi.org/10.3390/su11205729.

Laurenti, R., Singh, J., Frostell, B., Sinha, R., & Binder, C. (2018). The socio-economic embeddedness of the circular economy: an integrative framework. *Sustainability*, 10 (7), 2129. https://doi.org/10.3390/SU10072129.

Laurenti, R., Singh, J., Sinha, R., Potting, J., & Frostell, B. (2016). Unintended environmental consequences of improvement actions: a qualitative analysis of systems' structure and behavior. *Systems Research and Behavioral Science*, 33 (3), 381–399. https://doi.org/10.1002/sres.2330.

Laurenti, R., Sinha, R., Singh, J., & Frostell, B. (2015). Some pervasive challenges to sustainability by design of electronic products – a conceptual discussion. *Journal of Cleaner Production*, 108, 281–288. https://doi.org/10.1016/j.jclepro.2015.08.041.

Laurenti, R., Sinha, R., Singh, J., & Frostell, B. (2016). Towards addressing unintended environmental consequences: a planning framework. *Sustainable Development*, 24 (1), 1–17. https://doi.org/10.1002/sd.1601.

Mont, O., Plepys, A., Palgan, Y. V., Singh, J., Curtis, S., Zvolska, L., & Velez, A. M. A. (2019). *Urban Sharing in Amsterdam*. Lund University (Media-Tryck). https://portal. research.lu.se/portal/en/publications/urban-sharing-in-amsterdam (35e661f8–0fe2–49da-a820–8954e9793752).html.

Mont, O., Plepys, A., Voytenko Palgan, Y., Singh, J., Lehner, M., Curtis, S. K., Zvolska, L., & Velez, A. M. A. (2020). *Urban Sharing in Toronto City report no 2*. https://static1.squarespace.com/static/581097b4e3df28ce37b24947/t/ 5f0c1d23792782130c6c4883/1594629426451/Toronto+report_FINAL_web.pdf.

Palgan, Y. V., Mccormick, K., Leire, C., & Singh, J. (2019). Mobile lab on sharing in Malmö. www.sharingcities.se.

Plepys, A., & Singh, J. (2019). Challenges and research needs when evaluating sustainability impacts of alternative consumption models (in press). In O. Mont (Ed.), *Research Agenda for Sustainable Consumption Governance*. Edward Elgar Publishing.

Singh, J. (2014). *Towards a Sustainable Resource Management: A Broader Systems Approach to Product Design and Waste Management*. KTH Royal Institute of Technology. http://urn.kb.se/resolve?urn=urn:nbn:se:kth:diva-141126.

Singh, J. (2016). *Beyond Waste Management: Challenges to Sustainable Global Physical Resource Management* (PhD Thesis, KTH Royal Institute of Technology). http://urn.kb.se/resolve?urn=urn:nbn:se:kth:diva-186517.

Singh, J., & Arora, C. (2021). Upcycling, jugaad and repair cafes for prosumption. In K. Sung, J. Singh, & B. Bridgens (Eds.), *State-of-the-Art Upcycling Research and Practice*. Springer International Publishing. https://doi.org/10.1007/978-3-030-72640-9.

Singh, J., & Cooper, T. (2017). Towards a sustainable business model for *plastic shopping bag management in Sweden. Procedia CIRP*, 61, 679–684. https://doi.org/10.1016/j.procir.2016.11.268.

Singh, J., Cooper, T., Cole, C., Gnanapragasam, A., & Shapley, M. (2019). Evaluating approaches to resource management in consumer product sectors: an overview of global practices. *Journal of Cleaner Production*, 224, 218–237. https://doi.org/10.1016/J.JCLEPRO.2019.03.203.

Singh, J., Laurenti, R., Sinha, R., & Frostell, B. (2014). Progress and challenges to the global waste management system. *Waste Management & Research*, 32 (9), 800–812. https://doi.org/10.1177/0734242X14537868.

Singh, J., Mont, O., Winslow, J., Lehner, M., & Voytenko Palgan, Y. (2021). *Exploring social, economic and environmental consequences of collaborative production: the case of bike repair maker spaces in three European countries.* In N. F. Nissen & M. Jaeger-Erben (Eds.), PLATE – Product lifetimes and the environment, 3rd PLATE Conference, September 18–20, 2019Berlin, Germany (pp. 717–723). https://doi.org/10.14279/depositonce-9253.

Singh, J., & Ordoñez, I. (2016). Resource recovery from post-consumer waste: important lessons for the upcoming circular economy. *Journal of Cleaner Production*, 134, A, 342–353. https://doi.org/10.1016/j.jclepro.2015.12.020.

Singh, J., & Sung, K. (2021). Systems approach to scaling up global upcycling: framework for empirical research. In K. Sung, J. Singh, & B. Bridgens (Eds.), *State-of-the-Art Upcycling Research and Practice*. Springer International Publishing. https://doi.org/10.1007/978-3-030-72640-9.

Singh, J., Sung, K., Cooper, T., West, K., & Mont, O. (2019). Challenges and opportunities for scaling up upcycling businesses – the case of textile and wood upcycling businesses in the UK. *Resources, Conservation and Recycling*, 150, 104439. https://doi.org/10.1016/J.RESCONREC.2019.104439.

Sinha, R., Laurenti, R., Singh, J., & Frostell, B. M. (2020). An element flow analysis approach to support the circular economy. In M. Brandão, D. Lazarevic, & G. Finnveden (Eds.), *Handbook of the Circular Economy* (pp. 99–115). Edward Elgar Publishing. https://doi.org/10.4337/9781788972727.00017.

Sinha, R., Laurenti, R., Singh, J., Malmström, M. E., & Frostell, B. (2016). Identifying ways of closing the metal flow loop in the global mobile phone product system: a system dynamics modeling approach. *Resources, Conservation and Recycling*, 113, 65–76. https://doi.org/10.1016/j.resconrec.2016.05.010.

Sung, K., Cooper, T., Oehlmann, J., Singh, J., & Mont, O. (2020). Multi-stakeholder perspectives on scaling up UK fashion upcycling businesses. *Fashion Practice*, 12 (3), 331–350. https://doi.org/10.1080/17569370.2019.1701398.

Sung, K., Singh, J., & Bridgens, B. (2021). *State-of-the-Art Upcycling Research and Practice*. Springer International Publishing. https://doi.org/10.1007/978-3-030-72640-9.

3

NET ZERO

My drive for innovation, enterprise and lifelong learning

Abhishek Tiwary

The early years

I am an Xennial, which, according to the *Oxford Dictionary of English*, is "a person born between the late 1970s and early 1980s, after (or towards the end of) Generation X and before (or at the beginning of) the millennial" (OED, 2023). Interestingly, Xennials are described as having had an 'analogue' childhood and 'digital' young adulthood. In short – we have seen it all developing in front of our eyes! While preparing for my higher education, the notions of environmental impact and climate change were gaining strength. Perhaps at that time, given the know-how that already existed in terms of the ozone hole crisis following the historic United Nations treaty – the Montreal Protocol – to ban chlorofluorocarbons, or CFCs, in 1989 (UN, 1987), much of our attention was on making fridges environmentally friendly. I do not recall anyone talking of cars and power stations causing any concerns, and they were considered symbols of growth and prosperity at the time. However, soon after, the scene started to change. In 1992, the United Nations Framework Convention on Climate Change (UNFCCC) took shape, which recognised the "dangerous human interference with the climate system" and committed the state parties to reduce greenhouse gas emissions based on the scientific consensus that global warming is occurring and that human-made CO_2 emissions are driving it (UNFCCC, 1992). Until then, Mother Earth was considered the provider of the bountiful natural resources for our needs, and every nation was rampantly plundering them to fulfil their technological aspirations. Then came the Kyoto Protocol in 1997, the first international treaty to set legally binding targets to cut greenhouse gas emissions, which, despite all its limitations, started the bandwagon of setting a global agenda to combat climate change. For the first time, the following seven greenhouse gases were identified as the

DOI: 10.4324/9781003380566-5

main culprits on record: carbon dioxide (CO_2), methane (CH_4), nitrous oxide (N_2O), hydrofluorocarbons (HFCs), perfluorocarbons (PFCs), sulphur hexafluoride (SF_6), and nitrogen trifluoride (NF_3) (UNFCCC, 1997).

Curiously, this was about the time when students like me started paying greater attention to topics on environmental issues in the press and the media. On the one hand, this led to greater awareness of the state of the Earth's wellbeing (including the wellbeing of her most ingenious inhabitants to date – humans), and on the other hand, it led to heated discussion over Global North and Global South, respectively representing the categorising of countries with haves and have-nots in terms of the technologies to deliver net zero. I vividly recall my time spent in the British library in those days, going through the books and reports presenting some of the initial concepts and frameworks to deal with the mammoth challenges that humanity has created for itself. These incidents motivated me to take on formal education in environmental assessment and management, cross-cutting conventional boundaries of science and engineering degrees.

The research training years

My PhD research led me to try my modelling and data analysis skills, utilising engineering and geo-spatial analysis approaches. Fortunately, I had one full year of training across four departments/schools at the University of Nottingham – Engineering, Business, Life Sciences and Geography. This further enriched my cross-disciplinary training, enabling me to develop a systems perspective of the research problems ever since. I conducted my PhD research in developing a numerical model (in those days, Fortran was the king!) and eventually validated that using real-world field data, standing in open fields at the University of Nottingham's rural campus. I think this exercise of spending two plus years in the fields gave me a genuine admiration of how bespoke technology can be applied to solve the problems of the natural world and, in turn, how the natural world can inspire research in designing sustainable engineering solutions for the industrial world.

During the final stage of my PhD, I won a *Universitas21* scholarship to spend a term at the University of Virginia, USA. This opened my eyes to the level of passion young scientists have in the USA to work at remote field stations, often working through the night under harsh conditions to complete primary data collection. I built a connection with some of the faculty members at the university. During one of my post-doctoral research assignments later I collaborated well with one of them, who had moved to Penn State University. This was when I started considering integrating natural systems as a sustainable solution to mitigate climate change issues.

Since completing my PhD, I have been lucky to be part of all three rounds of the UK Engineering and Physical Sciences Research Councils (EPSRC) Sustainable Urban Environment (SUE) projects, which enabled me to work in large interdisciplinary consortia. It allowed me to interact with senior researchers across different groups at several UK universities. This comprised several desk-based and field studies as part of these multidisciplinary research teams, which has built my experience at the energy-environment nexus, mainly focusing on developing and evaluating parsimonious technologies to reduce energy and resource demand. I was fortunate to have my PhD supervisor, Prof Jeremy Colls, lending his experience in writing an entire book on air pollution, which has been well received as a one-stop reference by students and early career researchers on diverse topics related to air pollution research (Colls & Tiwary, 2009; Tiwary, Colls & Williams, 2018).

The future research leader

Progressively, I have applied multi-disciplinary, techno-centric and public-good-generating research. Some recent works have considered a win-win for carbon and pollution (air/water/residue) management in the urban environment and evaluation of policy measures. They have considered designing and evaluating a mix of technology and nature-based solutions with a well-rounded purpose, i.e., accelerating air and water pollution reduction potential in urban environments from measures implemented with the primary objective of climate change mitigation. My research contributions so far have been centred on sustainability – both at the process and systems level. When I take stock of all my research to date to consider as a future research leader, I find my niche in developing pollution control solutions (both air and water) with due consideration to energy/resource efficiency and waste reduction/reuse. I would see this as an instrumental part of delivering net zero in the near future while simultaneously safeguarding human health and wellbeing.

I have been involved in numerous funded projects broadly themed under sustainable/systems engineering, evaluating the feasibility of environmentally benign solutions to global challenges on pollution control, resource reuse and renewable energy for both industrial and societal applications. While most of my early year research centred mainly on sustainability themes pertaining to the urban environment, cross-cutting the air-water nexus, I have started appreciating the synergy towards net zero lately. For example, as a Researcher co-investigator on the EPSRC-SUE3 supported Self Conserving Urban Environment (SECURE) consortium project, I contributed to the concept development of reducing the carbon burden of urban systems by integrating activities across the transport, food, energy and waste sectors. Following this, I acquired a Marie Curie outgoing fellowship on the Bio-energy Technology Assessment – Environmental Burden Minimisation project (BETA-EBM), which funded my research in India for two years, working in the Energy and Environment Technology division of the Energy and

Resources Institute (TERI) in New Delhi. This phase allowed me to experience some innovative, low-cost, off-grid (i.e., without needing any dedicated electricity network to power electrical appliances) renewable energy technology. I spent considerable time and effort in making sustainable reuse of the waste by-products arising from biogas plants, which otherwise used to be dumped into landfill by default. I would consider this my stepping stone to the broader 'circular economy' concept.

Since 2019, I have been supported by the UK Royal Academy of Engineering and the British Council to evaluate the feasibility of integrating renewable energy and alternative treatment methods in improving the resource and energy sustainability of the wastewater treatment sector. Internationally, I have been involved in several collaborative projects (Jordan × 2, Nigeria and Togo, Turkey, Nepal, Qatar, UAE), mainly contributing to feasibility evaluations of innovative energy and pollution control integration using autonomous solutions. This has only been possible through my extensive collaboration on complementary aspects of project delivery, including academic and industrial collaborators with expertise in chemical engineering, pharmaceutical sciences, entrepreneurship, electrical engineering, etc. I am privileged to have forged strong international collaborations, which have resulted in the successful and timely delivery of all the projects. The industry–academia partnership with Jordan has also received recognition from the UK government regarding the contribution made to addressing a burning issue of resolving the water and wastewater crisis looming in Jordan's arid regions utilising a technology that can be entirely operated off-grid using renewable solar energy.

Since COVID-19, I have been involved in developing an integrated solution for a pathogen-free and energy-efficient warming/cooling system for liveable indoor space. This supports the UK government's strategic priority areas of clean air and low carbon heating. Through this, I have won a Royal Academy of Engineering Industry Fellowship titled Enabling Hybrid Autonomous Non-conventional System for Clean Indoor Environment (ENHANCE) project (www.enhanceprojects.info). I have been selected to display a demonstrator hybrid indoor air cleaning/warming unit at the British Science Festival Leicester Event (September 2022). This has been developed via the Innovate UK Small Business Research Initiative, partnering with a UK industry.

Net zero policies and control measures enacted to mitigate climate change present opportunities and threats to broader environmental sustainability. I have developed an integrated assessment framework combining the contributions of mobility and energy sector technologies evaluated, incorporating some emerging trends to ascertain their true sustainability potential (Figure 3.1).

An example assessment evaluating the pros and cons of co-management (i.e., concerted actions towards climate change and air quality management) through local sustainability initiatives found that interventions to reduce carbon emissions from the mobility and energy sector could inadvertently

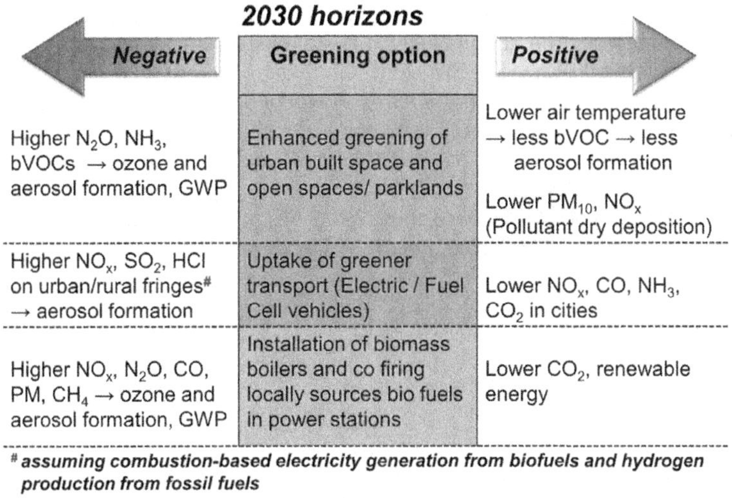

2030 horizons

Negative	Greening option	Positive
Higher N_2O, NH_3, bVOCs $\rightarrow$ ozone and aerosol formation, GWP	Enhanced greening of urban built space and open spaces/ parklands	Lower air temperature $\rightarrow$ less bVOC $\rightarrow$ less aerosol formation Lower PM_{10}, NO_x (Pollutant dry deposition)
Higher NO_x, SO_2, HCl on urban/rural fringes[#] $\rightarrow$ aerosol formation	Uptake of greener transport (Electric / Fuel Cell vehicles)	Lower NO_x, CO, NH_3, CO_2 in cities
Higher NO_x, N_2O, CO, PM, CH_4 $\rightarrow$ ozone and aerosol formation, GWP	Installation of biomass boilers and co firing locally sources bio fuels in power stations	Lower CO_2, renewable energy

[#] *assuming combustion-based electricity generation from biofuels and hydrogen production from fossil fuels*

FIGURE 3.1 Systems scale assessment framework evaluating the interactions of anthropogenic and biogenic emissions from plausible greening initiatives in three strands – energy, mobility, ecosystem

Source: Tiwary et al. (2013)

impact the local and regional environment, specifically on air quality, if not thought through. Potential implementation of this framework in evaluating systems-scale interactions of net zero interventions at urban scale (e.g., green mobility, renewable energy and green infrastructure) was discussed. This calls for a step-change through more cohesive, cross-disciplinary policy frameworks at the energy–environment nexus, going beyond the local administrative spheres to maximise the 'win-win' potential for achieving sustainable net zero.

Going forward, crucial to the multidisciplinary ethos of the net zero theme, I plan to build on my longitudinal research track record in national and international projects to develop resource-efficient, zero-pollution systems at various scales (micro-to-urban) within my area at the nexus of pollution control and carbon management. I am prepared to take this as my lifelong learning mission.

References

Colls, J., & Tiwary, A. (2009). *Air pollution: measurement, modelling and mitigation*. CRC Press.

OED (2023). *Updates to the OED new words list December 2021*. Oxford English Dictionary. Oxford University Press. December 2021. Retrieved 17 March 2023.

Tiwary, A., Colls, J. J. & Williams, I. (2018). *Air pollution: measurement, modelling and mitigation*. CRC Press.

Tiwary, A., Namdeo, A., Fuentes, J., Dore, A., Hu, X. M., & Bell, M. (2013). Systems scale assessment of the sustainability implications of emerging green initiatives. *Environmental Pollution*, 183, 213–223.

UN (1987). *The Ozone Hole – the Montreal Protocol on Substances that Deplete the Ozone Layer*. Theozonehole.com. 16 September 1987.

UNFCCC (1992). *Article 2 – the United Nations Framework Convention on Climate Change*. The United Nations.

UNFCCC (1997). *Kyoto Protocol to the United Nations Framework Convention on Climate Change at COP3*. Kyoto, Japan. 11 December 1997.

PART II
DESIGN AND INNOVATION

4

DRIVING SUSTAINABILITY AND INNOVATION THROUGH DESIGN

The BMW Group + QUT Design Academy

Rafael Gomez

Introduction

BMW Group is a sustainability leader in the automotive world and has been consistently named at the top of the S&P Dow Jones Sustainable Index rankings for over two decades (BMW Group, 2020). BMW Group has positioned sustainability as core to its future viability and has committed to ambitious targets with a goal of total climate neutrality by 2050 at the latest. In 2016, BMW Group set its vision for the next century, identifying six megatrends that will shape mobility: versatility, connectedness, tailor-made, humanised, emission-free energy, and diversity (BMW Group, 2016). These serve as the foundation embodying BMW Group's vision for the future while highlighting the importance of creativity, digitalisation, flexibility, a human-centred approach, and sustainability for the company. These also closely mirror the skillsets and capabilities embodied by design. Design is a creative discipline with a human-centred approach with sustainability in mind. Designers have a particular skillset with digital tools and a flexible mindset that supports innovation. For these reasons, designers' capabilities and attributes have been a valuable match for BMW Group – and this is why Queensland University of Technology (QUT) School of Design has established a strong relationship with the company. With its focus on creativity and innovation, design has become a crucial part of the projects we work on with BMW Group, with digital skills being the enabler for this compatibility. Digitalisation in the context of BMW Group includes digital twins, visualisation, data, technologies, and workflows used for cutting-edge projects for implementation across design, production, manufacturing, and logistics.

To capitalise on this opportunity, I planned to connect design expertise with BMW Group's future strategy. This evolved into the BMW Group +

DOI: 10.4324/9781003380566-7

QUT Design Academy (the Academy) that I founded with a colleague, Jimmy Nassif, who worked at BMW Group at the time. The Academy is based on three pillars: Internships, Research, and Special Projects, which work together to drive positive outcomes for BMW Group and QUT. The Academy now plays a vital role by embedding designers in traditionally non-design roles and divisions to help explore how new and emerging digital technologies can be leveraged in achieving BMW's future ambitions.

In this chapter, I provide an overview of the unique partnership between BMW Group and QUT, a brief history of the relationship, the relevance of design, how we work together to achieve sustainability goals with project examples, discuss the impacts of our collaboration on sustainability, and conclude with future partnership directions. The information in this chapter helps inform readers about various elements regarding academia, research, and design. First, it sheds light on the unique collaboration between QUT and BMW Group and a way to structure the partnership that provides positive outcomes for academic and industry goals. Second, it illustrates how design can play a critical role in contributing to the ambitious sustainability goals of a global corporation through creativity, innovation, and digital expertise. Third, it demonstrates examples of cutting-edge projects we have engaged in and the power emerging digital technologies, such as digital twins, play in shaping a sustainable future in the global automotive sector. Fourth, it inspires other design researchers and academics to find novel ways to engage with leading multinational corporations in the context of sustainability, given the complex environmental challenges we face now and in the future.

BMW Group + QUT Design Academy

The BMW Group + QUT Design Academy is the culmination of a long-term collaboration between the BMW Group in Germany and the Queensland University of Technology, School of Design in Australia. BMW Group (BMW, Rolls Royce, MINI, BMW Motorrad, and idealworks) is a Global Fortune 100 company. It is the world's leading manufacturer of premium automobiles and motorcycles headquartered in Munich, Germany, with a significant global reach (BMW Group, 2023a). Queensland University of Technology is an ambitious educational institution with a growing research output focused on technology and innovation. It promotes itself as 'the university for the real world' with a clear focus on conducting research and education relevant to industry, communities, and society, offering academic programmes across five faculties (QUT, 2023). Industrial Design, within the School of Design, is a discipline focusing on creating products, services, and systems to improve people's everyday lives, positively impact the environment, and tackle complex problems through visionary and creative thinking.

Brief history

The BMW Group + QUT Design Academy journey began in 2016 when I facilitated the first paid internship for an Industrial Design student at BMW Group in Munich, Germany. We established a unique partnership between the two organisations through strategic efforts and networking. In 2018, BMW Group signed an International Collaborative Agreement with QUT, and in 2019, a Statement of Intent was signed to establish the BMW Group + QUT Design Academy. The primary objective was leveraging design expertise to advance innovation within the mobility sector. In 2020, the Academy was officially launched at QUT in Brisbane, serving as a hub for design and research collaboration with BMW Group.

The accomplishments of our collaboration are noteworthy. Our inaugural intern, Dylan Sheppard, successfully signed a lifetime contract with BMW Group and became the Digital Design Lead at idealworks (a BMW Group subsidiary). We have secured paid internships for more than 30 students at BMW Group in Germany and the UK. Additionally, the research programme has seen the involvement of one MPhil student (completed) and three PhD students. BMW Group also provided a BMW M8 Competition Coupe supercar for teaching and research purposes. Currently, eight design and research associates contribute to the Academy's initiatives. Furthermore, the Academy team has worked on various projects with significant impacts, contributing to establishing idealworks, assisting the flagship BMW iFACTORY project, and facilitating the publication of SORDI, the world's largest open-source dataset for artificial intelligence applications in production.

As the founder and lead of the partnership, I manage all aspects of the Academy's operations and work alongside a team of designers and research associates striving to achieve the objectives set forth by the partnership between BMW Group and QUT.

The BMW Group + QUT Design Academy structure

The Academy is a design hub for advanced learning, cutting-edge research, and Research and Development projects focused on the application of emerging technologies, including virtual reality (VR), augmented reality (AR), artificial intelligence (AI), autonomous systems, and robotics. I developed a structure based on three pillars that would serve QUT's and BMW Group's mission: **Internships, Research**, and **Special Projects**. The Academy quickly evolved into a bespoke teaching and research workspace that provides real-world opportunities for design students, graduates, and researchers to hone their skills, advance their learning, and work on innovative projects with BMW Group. At the same time, it assists BMW Group in staying at the forefront of innovation and sustainability through the application of design and emerging technologies.

Pillar 1: Internships. Offers high-performing QUT design students an opportunity to advance their learning through real-world projects. It provides a pathway for paid internship placements at BMW Group in Germany. Students progress through a six-level advancement track, with distinct objectives and responsibilities that promote the necessary technical and personal capabilities to enhance their professional development.

Pillar 2: Research. A progressive research agenda for PhD and MPhil students to conduct world-class research intersecting with BMW Group and QUT's expertise. This includes design thinking, emerging technologies, industry 4.0, visionary human–machine interface and related fields.

Pillar 3: Special Projects. Projects tailored toward professional designers established by BMW Group in Munich and administered by the Academy team, working with various departments, divisions, and subsidiaries across the BMW Group network.

The dynamic of these three pillars is vital, with each supporting and contributing to the others (Figure 4.1). The Internship pillar helps cultivate an adequate pool of students who learn skills relevant to BMW Group and those needed for the Special Projects pillar. As students complete internships at BMW Group, they pave the way for future project partnerships. Likewise, interns might be interested in pursuing MPhils and PhDs on their return. Another example is the Special Projects pillar's impact on Internships and Research. As we conduct more real-world projects with BMW Group, our reputation increases, attracting more students.

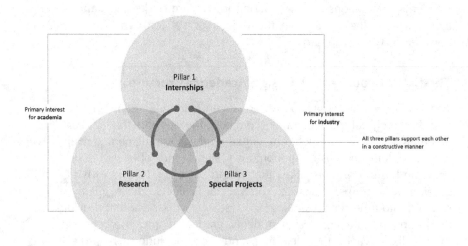

FIGURE 4.1 The BMW Group + QUT Design Academy three-pillar structure. Image created by the author

It is also relevant to highlight that this structure serves important academic goals through research and internships, while industry needs are addressed primarily through Special Projects and Internships. To develop trust with industry partners, it is essential to serve their needs, not just academic ambitions. Academic goals should be addressed. However, to achieve some wins for the industry partner, it is crucial to gain traction and build meaningful trusting relationships with industry partners (Zelenko, Gomez, & Kelly, 2021).

The Academy has two dedicated spaces to conduct its activities: Alpha Studio and Beta Lab. Alpha Studio is the centre of all Academy operations and is home to our Design Associates working on Special Projects alongside the Internship programme for QUT students. Beta Lab is a design research, teaching, and presentation space. With exclusive access to a BMW M8 Competition Coupe vehicle, research associates can explore cutting-edge topics, including human–machine interfaces and interactive automotive technologies. The vehicle facilitates learning for undergraduate students through projects embedded in the design degree. We also have access to numerous pieces of design and technical equipment to assist with the various projects.

Why design is relevant for BMW Group

In 2016, the company celebrated its centenary and set its sights on the future by celebrating with the motto 'THE NEXT 100 YEARS'. The company formulated six megatrends (BMW Group, 2016) that would impact the coming decades of mobility, including:

i **Versatility**: New forms of mobility will open countless possibilities for people to get where they want to go.
ii **Connectedness**: Digitalisation and digital intelligence are meant to serve people; this is the only way to permanently enhance people's quality of life.
iii **Tailor-made**: Mobility will be increasingly flexible and tailored to individual needs, with customisation ensuring that people can use the best means of transport and take their preferred route to their destination.
iv **Humanised**: Technologies are getting smarter, and innovations are only beneficial to humans if they are simple and user-friendly. Technologies must learn from and adapt to people, making them appear less technical and more human and familiar.
v **Emission-free energy**: Energy will increasingly come from renewable sources with a clear need for environmentally compatible vehicles built using renewable energies and recycled without generating emissions.
vi **Diverse**: It will become even more critical for global companies to take responsibility for the environment and the people directly or indirectly in their sphere of activity.

These megatrends highlight the importance of creativity, digitalisation, flexibility, a human-centred approach, and sustainability for BMW Group. These are all skillsets and capabilities that are central to the design. Further, designers have a particular skillset in digital tools and a flexible mindset that allows them to be innovative given any challenge. For these reasons, designers' skills, capabilities, and attributes cultivated through their education, facilitated by the Academy, positions them ideally to contribute to the goals and objectives of BMW Group.

Sustainability at BMW Group

Sustainability has become central to BMW Group's strategic direction, from the supply chain to production to the end of the use phase of all products (BMW Group, 2023b). For over two decades, BMW Group has led the way in sustainability in the automotive industry. They have taken a holistic approach to the entire lifecycle across the whole value chain from raw material extraction and selection, supply chain management, production, manufacturing, logistics, and final product to end-of-life reclaiming and re-use of materials. According to the S&P Dow Jones Sustainability Index (the most extended running index of its kind), BMW Group is the only automobile manufacturer consistently named at the top of these rankings, including in the latest assessment conducted in 2020 (BMW Group, 2020).

BMW Group's declared goals for sustainability centre on six critical criteria driving a holistic approach to sustainability: (i) CO_2 reduction; (ii) Electromobility; (iii) Circularity; (iv) Environmental and Social Standards; (v) Employees; and (vi) Society. The CO_2 reduction target spans the entire lifecycle, from the supply chain through production to the end of the use phase, intending to be climate-neutral no later than 2050; an expansion of electromobility with a goal of ten million fully electric vehicles on the roads by 2030; circularity through vehicles that are developed with components that can be fully recovered and re-used to avoid consuming primary materials and committed to the highest environmental and social standards internally and with suppliers worldwide. It values employees by providing an environment that authentically reflects their highest regard. As a global player, the company is aware of its responsibility towards society with targeted sustainable, social, and intercultural commitments designed to help people and cultures globally (BMW Group, 2023b).

A prominent example of this approach is captured in BMW's i Vision Circular concept (BMW Group, 2023c). The concept is conceived in line with the principles of circularity and optimised for closed material cycles through the entire design, development, and manufacture process. The goal is to achieve 100% recycled material use and 100% recyclability based on circular design principles. Some innovations include significantly reducing parts and materials, an exterior design with almost 100% recycled mono materials (instead of composites), and all materials being returned directly to material cycles. The interior is also made from 100% recycled mono materials, able to be returned to material cycles. Another innovative feature is the BMW logo engraved into the hood instead of being added as a separate badge, minimising materials, time, and embedded energy. This might seem like a minor change, but this has a significant impact when multiplied across the entire sum of cars produced by BMW.

This reflects BMW Group's considerable efforts to position sustainability at the centre of its future viability. BMW Group has developed into a sustainability leader through a thoughtful and sophisticated strategy that sets trends in production technology and sustainability with a shift towards digitalisation and efficient production.

The BMW iFACTORY: sustainability through digitalisation

The BMW iFACTORY drives BMW Group's sustainability-focused transformation. A flagship project revolutionising automotive production worldwide and accelerating sustainable e-mobility, the iFACTORY is guided by three core concepts: LEAN, GREEN, and DIGITAL (BMW Group, 2023d).

i **LEAN** focuses on efficient and adaptable manufacturing, integrating and enhancing the variability of processes. This allows the production of diverse vehicle types on a single line, resulting in increased agility, streamlined procedures, and more competitive production.

ii **GREEN** is dedicated to advancing sustainable production practices. BMW Group facilities globally procure green electricity exclusively and adhere to circular economy principles during production.

iii **DIGITAL** strategically leverages cutting-edge innovations such as virtualisation, AI, and data science to seamlessly connect all aspects of automotive production. This enables heightened data transparency, novel networking approaches, optimal data utilisation, and improved efficiency and effectiveness in operations.

The implementation of iFACTORY enables sustainable production through digitalisation. BMW Group utilises virtualisation, AI, and data to network various aspects of automotive production. By fully capturing production sites in a 3D scan, planning work can be conducted remotely, anytime and anywhere. This digital twin technology allows real-time virtual walkthroughs across different locations and time zones. Employees can immerse themselves in a digital workplace representation, capture insights, and improve. The iFACTORY demonstrates its commitment to sustainable e-mobility and driving innovation in the automotive industry.

Recent studies have pointed towards the benefits, limitations, and challenges of digitalisation for sustainability objectives (Burinskienė & Seržantė,

2022; Cricelli & Strazzullo, 2021; Renn, Beier, & Schweizer, 2021; Rasheed, San, & Kvamsdal, 2020; Shah, Menon, Ojo, & Ganji, 2020), and for the iFACTORY project digital twins form a critical part of this agenda. Digital twins, defined as a virtual representation of objects, process, service, or environment that behaves and looks like their counterparts in the real world (Grieves, 2016), are utilised to test, plan, manage, monitor, and make decisions on all aspects of what is being digitally reproduced. There is a recognition that digital twins can contribute to broad sustainability goals, including efficient resource management, safe innovations in green technologies, inclusive partnerships for sustainability, and monitoring the progress of sustainability goals (Tzachor, Sabri, Richards, Rajabifard, & Acuto, 2022). In the context of manufacturing, digital twins offer the potential for improved product quality, higher productivity, lower cost, and increased manufacturing flexibility, resulting in smarter, more efficient, and more effective manufacturing (He & Bai, 2021), ultimately improving sustainability outcomes.

Digital twins have immense potential in manufacturing. Companies can test and plan multiple factors simultaneously and in real time by creating a virtual factory replica. Managers can explore different designs and asset distributions, engineers can optimise engineering systems, and plant staff can practice working with new equipment virtually. Autonomous Mobile Robots (AMRs) can also be trained in hyper-realistic virtual environments to detect objects and routes, minimising risks before deployment. This concurrent collaboration in the digital twin environment provides real-time feedback to stakeholders, aiding critical decision-making.

BMW Group is at the forefront of this innovation, applying digital twins in their Debrecen plant in Hungary more than two years before they begin production of the plant (Malayil, 2023). This pioneering approach allows for complete virtual planning and validation of the facility's layout, robotics, and logistic systems. It signifies BMW's transition to the iFACTORY strategy and establishes a fully-fledged industrial metaverse application (Kshetri, 2023). The integration of digital twin technology and the emphasis on design highlight BMW Group's commitment to pushing the boundaries of manufacturing and sustainability, setting the stage for future advancements in the automotive industry.

As the company transitioned to the iFACTORY strategy, it recognised the decisive role that innovation and creativity play in this context. This is why design began to play an important role. As various divisions across BMW Group explored how to achieve its ambitious digital and sustainability goals, QUT interns – with an educational background founded on creativity, digital capabilities, and visualisation skillsets – began to apply design skillsets to positively impact the innovative projects they were working on.

Digitalisation expertise: how we help BMW achieve sustainability goals

Our expertise in design, digital technologies, digital visualisation, digital workflows, and specialised software knowledge has allowed us to positively contribute to BMW Group's sustainability objectives through the iFACTORY initiative. I would like to present four projects in which our design students and associates have been involved. This includes **3D Asset Creation**, the **SORDI** project, **Accelerating Creation of Digital Twins**, and **Green Physics AI**. Each example outlines the project, its benefits to BMW Group's mission for sustainability through digitalisation, and how we have contributed through design expertise.

3D asset creation and library

We are currently assisting BMW Group with a critical project involving the creation of 3D asset libraries (Figure 4.2, top left). This is part of their mission to create digital twins of all their factories, spaces, and objects, supporting current and future logistics planning efforts. 3D asset creation involves generating highly realistic digital replicas of all factory objects. This is the basis of developing digital twins in their factories.

One specific way this helps is to train AI systems in 1:1 scale digital environment without disrupting production on the factory floor. The training data gained in the digital environment is implemented in a fleet of AMRs in

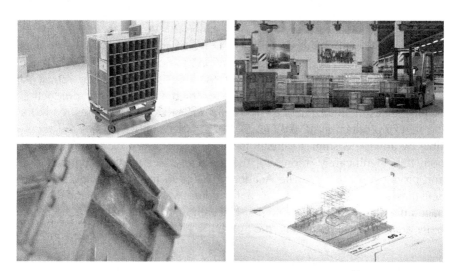

FIGURE 4.2 Examples of virtual factory trolley (top left), virtual factory asset class (top right), a virtual container with wear and tear (bottom left), and virtual AI point cloud workflow animation still. All images were created by BMW Group + QUT Design Academy

the real world. Central to training AI systems is object recognition. In short, this means that engineers can instruct the robot whether an object is a type of vehicle, person, container, and so on and how it should interact with another. The robot continues to develop its object recognition capability as it is trained within the digital twin, which employs highly realistic 3D assets. Once the robots can identify different objects, they can make their own decisions when navigating spaces and performing tasks in the real world.

The Academy's involvement in this project has led to creating an extensive, highly realistic 3D asset library for software engineers at BMW Group and idealworks. Our team have modelled and visualised well over 200 factory assets for BMW Group, which have been consolidated into an online library accessible to BMW Group and idealworks staff globally. All assets comply with industry standards and contribute to automation, AI, smart machines, smart vehicles, and smart factory developments as part of the iFACTORY strategy.

SORDI

Since 2022, we have supported idealworks and BMW Group's TechOffice on the Synthetic Object Recognition Dataset for Industries project (SORDI.ai). The project is a collaboration between BMW Group TechOffice Munich, idealworks, NVIDIA, and Microsoft that enables software developers to utilise industry standard datasets to train computer vision models (Abou Akar, Tekli, Jess, Khoury, Kamradt, & Guthe, 2022). SORDI.ai is a synthesised dataset containing 80 object classes and over 800,000 photorealistic images in various modalities designed to suit general image processing tasks, including classification, object detection, or segmentation (BMW Group, 2022).

Our contribution included creating a range of assets across various object classes, including Logistics, Transportation, Office Settings, and Signage (Figure 4.2, top right). IT experts can use the assets to create robust AI models to develop tailored solutions for manufacturing and production employees to maintain mature AI systems for validation purposes ready for the start of production (BMW Group, 2022). These object classes are accessible by anyone wishing to streamline and accelerate AI training in production, thus significantly improving the workflows within the manufacturing and production contexts. The Academy has been involved in several other activities for the SORDI.ai project, including website renders and landing page animation, the 2022 SORDI.ai Hackathon promotional animation, and other animations that illustrate the SORDI.ai system.

Green Physics AI

Green Physics AI metadata aims to provide real-time insights into asset conditions and usage. Leveraging historical records, this metadata undergoes processing while advanced AI prediction models forecast assets' future status and potential risks (SORDI.ai, 2023). Critical parameters of industrial asset lifecycles can be simulated, such as deformations, wear and tear, contamination, and other relevant variables that may have occurred during the specified timeframe (Figure 4.2, bottom left). It accurately simulates the asset's condition over time through automatic calculations and offers predictions for its future status. Leveraging NVIDIA Omniverse (a real-time 3D graphics collaboration platform created by NVIDIA), it can simultaneously simulate thousands of assets and their interdependencies, with the ability to estimate the impact of small decisions on the overall system in the future. With this valuable information, BMW Group can effectively monitor and analyse asset performance, make informed decisions, mitigate risks, and optimise operations for optimal efficiency and asset utilisation. Green Physics AI metadata is essential in the digitalised decision-making process for efficient and sustainable industrial practices.

Green Physics AI metadata brings static assets to life. This means that the models are no longer limited to a single, unchanging digital texture (the materials and features of the assets in the digital environment) but are updated based on significant inputs and conditions. This makes these virtual assets interactive, providing a more realistic representation of the object over time to plan accordingly. We are currently working with BMW Group on developing this system and have assisted them in generating digital animations and visualisations to help explain this concept by simple, effective means. This helps explain the complex information in an accessible and engaging format.

Accelerating the creation of digital twins

BMW Group TechOffice in Munich has begun developing a new workflow that could accelerate the development of future digital twins through AI, data, and 3D scans (Figure 4.2, bottom right). It involves generating a point cloud of an existing space, then utilising AI for scene reconstruction, and assigning 3D assets based on this information into the environment. This process can recreate a highly realistic and valuable digital twin of the selected space, including the digital assets.

We are in the early stages of assisting BMW Group TechOffice on this project. Through our expertise in digital visualisation, we have helped the BMW Group teams communicate this new workflow that generates digital twins more rapidly than traditional techniques. AI is used for scene reconstruction in the digital twin by using existing object recognition methods and 6D pose estimation (the task of determining the six-degrees-of-freedom [6D] pose of an object in space based on red, green, and blue digital images). This allows the AI to determine various elements, such as the age of the asset if the containers are filled or empty, and how the object is arranged.

Summary and future work

BMW Group is renowned for its sustainability performance in the automotive industry, consistently ranking highly on sustainability indices. With a commitment to achieving climate neutrality by 2050, the company identified megatrends emphasising creativity, digitalisation, flexibility, human-centred design, and sustainability, which closely align with the design field. Recognising the value designers bring to these areas, the BMW Group + QUT Design Academy was established to foster collaboration between QUT and BMW Group to achieve ambitious sustainability goals. The Academy is centred on three pillars: Internships, Research, and Special Projects purposefully designed to achieve industry and academic aspirations.

The input of our BMW Group + QUT Design Academy students and associates has been significant – especially through the application of digital expertise. We have played a role in establishing new subsidiary companies for BMW Group that are involved in projects that have developed into flagship initiatives such as BMW iFACTORY, and continue to engage in activities that forge innovative solutions to achieve sustainability goals.

Specifically, with a unique blend of creativity and digital skills, the Academy designers and students have worked on various projects that assist BMW Group in becoming an industry leader in digitalisation and sustainability. We have generated digital asset libraries that contribute to developing fully realised digital twins of current and future factories. We have also contributed to creating SORDI.ai, enabling software developers to freely utilise industry-standard datasets to train computer vision models. Furthermore, we have supported Green Physics AI using metadata that can provide real-time insights into digital asset conditions and usage factors such as deformations, wear and tear, contamination, and other variables that may occur over the asset's life. This creates virtual assets that are more realistic representations of the object over time, allowing accurate digital twins. We are in the early phases of another innovative project with the potential to revolutionise digital twin creation through AI, data, and 3D scan workflows.

Applying unique capabilities, including creativity and digital skillsets, has enabled the Academy teams to positively contribute to these and other projects. The impact this is having on the teams we work with and on BMW Group's sustainability goals through the iFACTORY project is considerable and an example of the potential positive impact of design on a global scale.

The Academy is excited about the future partnership with BMW Group, and we will continue to explore projects that utilise digital design skills for future production, manufacturing, and logistics. Additionally, the Academy is involved in projects related to automotive infotainment innovations, human–machine interface designs, and autonomous robotic systems. The enduring partnership between the BMW Group and QUT is a source of pride. I am excited about designers' contributions and skillsets in driving sustainability goals for a global multinational corporation such as BMW Group.

Acknowledgements

Thank you to the BMW Group TechOffice and idealworks teams for their ongoing support, including Mark Kamradt and Chafic Abou Akar at BMW Group TechOffice and Jimmy Nassif and Dylan Sheppard at idealworks for their unwavering support. A special thanks to Professor Mark Harvey, Professor Scott Sheppard, Professor Mandy Thomas, Professor Lori Lockyer, Professor Lisa Scharoun, and all the professional and administrative staff at QUT for their incredible support, assistance, and encouragement over the years. Finally, I would like to thank and acknowledge the incredibly talented team at the BMW Group + QUT Design Academy, including Jordan Domjahn, Michael Williams, Timothy Lim, Epifanio Pereira, James Dwyer, Mike Lepre, and a host of other design associates, students, and colleagues who have made this initiative a success over the years.

References

Abou Akar, C., Tekli, J., Jess, D., Khoury, M., Kamradt, M., & Guthe, M. (2022). *Synthetic Object Recognition Dataset for Industries*. In 2022 35th SIBGRAPI Conference on Graphics, Patterns and Images (SIBGRAPI) (Vol. 1, pp. 150–155). IEEE.

Belfield, H. (2012). *Making Industry–University Partnerships Work: Lessons from Successful Collaborations*. Science/Business/innovation Board AISBL.

BMW Group. (2016). *BMW Group. The Next 100 Years*. https://www.press.bmwgroup.com/global/article/detail/T0261088EN/bmw-group-the-next-100-years?language=en.

BMW Group. (2020). *BMW Group Named Sector Leader in Dow Jones Sustainability Indices 2020*. https://www.press.bmwgroup.com/global/article/detail/T0321071EN/bmw-group-named-sector-leader-in-dow-jones-sustainability-indices-2020?language=en.

BMW Group. (2022). *BMW Group Publishes SORDI, the Largest Open-source Dataset by Far for Super-Efficient AI Applications in Production*. https://www.press.bmwgroup.com/global/article/detail/T0375993EN/bmw-group-publishes-sordi-the-largest-open-source-dataset-by-far-for-super-efficient-ai-applications-in-production?language=en.

BMW Group. (2023a). *Business Segments*. https://www.bmwgroup.com/en/company/business-segments.html.

BMW Group. (2023b). *Our Goals*. https://www.bmwgroup.com/en/sustainability/our-goals.html.

BMW Group. (2023c). *The BMW i Vision Circular*. https://www.bmw.com/en-au/discover/concept-vehicle/bmw-i-vision-circular-highlights.html.

BMW Group. (2023d). *This Is How Digital the BMW iFACTORY Is*. https://www.bmwgroup.com/en/news/general/2022/bmw-ifactory-digital.html.

Burinskien, A., & Seržant, M. (2022). Digitalisation as the Indicator of the Evidence of Sustainability in the European Union. *Sustainability*, 14 (14), 8371.

Cricelli, L., & Strazzullo, S. (2021). The Economic Aspect of Digital Sustainability: A Systematic Review. *Sustainability*, 13, 8241. https://doi.org/10.3390/su13158241.

Grieves, M. (2016). *Origins of the Digital Twin Concept*. doi:10.13140/RG.2.2.26367.61609.

He, B., & Bai, K. J. (2021). Digital Twin-based Sustainable Intelligent Manufacturing: A Review. *Advances in Manufacturing*, 9, 1–21.

Kshetri, N. (2023). The Economics of the Industrial Metaverse. *IT Professional*, 25(1), 84–88.

Malayil, J. (2023). *BMW Starts Virtual Production of Its Next-gen EVs Using NVIDIA's Omniverse*. https://interestingengineering.com/innovation/bmw-starts-virtual-production-of-its-next-gen-evs-using-nvidias-omniverse.

QUT. (2023). *Our University*. https://www.qut.edu.au/about/our-university.

Rasheed, A., San, O., & Kvamsdal, T. (2020). Digital Twin: Values, Challenges and Enablers from a Modeling Perspective. *IEEE Access*, 8, 21980–22012.

Renn, O., Beier, G., & Schweizer, P. J. (2021). The Opportunities and Risks of Digitalisation for Sustainable Development: A Systemic Perspective. *GAIA-Ecological Perspectives for Science and Society*, 30 (1), 23–28.

Shah, S., Menon, S., Ojo, O. O., & Ganji, E. N. (2020, November). *Digitalisation in Sustainable Manufacturing – a Literature Review*. In 2020 IEEE International Conference on Technology Management, Operations and Decisions (ICTMOD) (pp. 1–6). IEEE.

SORDI.ai. (2023). *SORDI.ai is now GREEN!*https://sordi.ai/green.

Tzachor, A., Sabri, S., Richards, C. E., Rajabifard, A., & Acuto, M. (2022). Potential and Limitations of Digital Twins to Achieve the Sustainable Development Goals. *Nature Sustainability*, 5 (10), 822–829.

Zelenko, O., Gomez, R., & Kelly, N. (2021). Research Co-Design: Meaningful Collaboration in Research. In Blackler, A., & Miller, E. (Eds.) *How to Be a Design Academic: From Learning to Leading*. CRC Press, pp. 227–243.

5

"I BOMME AS A BOMBYLL BEE DOTHE"

An (un)random research journey

Garrath T. Wilson

The designer

The first thing to say about this chapter is that it is very much aimed at early career researchers or those new to academia. I do not know how the reader will receive this recounted journey, but it is honest, and, hopefully, of interest. Incidentally, when I chose the title – "I bomme as a bombyll bee dothe" (Oxford English Dictionary, 2023) – I thought it was quite a clever way to convey that I perhaps 'bumble around', moving between projects as an academic, and maybe there is a little truth in that, as I will unpack within this chapter. I rapidly realised, however, that it was also quite an unwieldy title, and I regretted it immediately after I sent it off to the editors. The point is that what I am going to present here is my personal trajectory, which I have found to not always be a straight line; in my words, how I have gone from designer to early career design researcher through to today: an academic at Loughborough University. To give the full title, I am (as of early 2023) Senior Lecturer in Experience Design (experience design is not just looking at physical artefacts or digital artefacts, but it is looking at the combination of those and how human–technology interaction shapes lives, emotions, and experiences). I have been in academia since 2009, when I started my doctoral research. I was made a Lecturer in Industrial Design in 2015, and in 2021, I was promoted to my current post. However, for context, we need to go back to 1981.

Research journeys do not typically need to go back to the birth of the individual, and do not worry; this chapter will not cover every year of my life. This needs framing upfront and to be recognised, especially when anyone purports to come with any advice, is my privilege and background. I am a white male academic from the UK, based in the UK, and

DOI: 10.4324/9781003380566-8

predominately operating within the Global North. This has undeniably opened doors for me and removed specific barriers in my research career. This needs acknowledging from the start of this chapter, and I wish the reader to bear this in mind. I am aware that this journey would either have been radically different or snuffed out from the start for others. There are other voices and journeys within this book that may resonate much better than my own, and I would encourage you to seek them out. I also recommend the excellent *The Black Experience in Design* (Berry et al., 2022). My journey has been transitioning from a working-class family background to acceptance within the distinctly middle-class UK academy. My parents left school at 14 to go and work in factories, so they had a 'job for life' mentality. Although supportive of education, I am not entirely sure if my parents knew what to do with it other than that it could bring better opportunities, the "opportunities we never had". However, that also comes with a hefty dose of pressure to succeed.

Furthermore, by success, I mean – in their eyes – to have a choice. It was never about not working in a factory but being able to choose not to. Despite their support, the more I succeeded, progressing through first a design career and then an academic career, the less relatable I became. At the start of my academic career in particular, I found it quite difficult to discuss what I did; my parents usually pulled out the phrase, "your problem is that you overthink everything", but then equally, I would go to the office, and I would have to rehearse in my head the correct 'intelligent' language to use. I would find myself over-planning my actions not to give away my more profound feelings of not quite belonging. Even today, I have a middle-class academic imposter syndrome, which I think frames many of the decisions I have made in my career. However, to quote my mum, "All you can do is try your best".

This is how I have never had a five-year or a ten-year plan, or generally even a next-year plan until recently, because that is not typical 'job for life' thinking. However, I was encouraged to seek new opportunities and 'try my best', so I did that. I enjoyed art and design, so I tried my best at art and design, and was quite good at it. I wound up studying Product Design (BSc) at the University of Central Lancashire, where I graduated with a first-class degree; then – after a short placement at DCA Design International and a fleeting period working as a design engineer on London Underground train door assemblies – I completed the Industrial Design (MSc) programme at Loughborough University, achieving a distinction. However, it was not all good times, and I had a few false starts at the beginning. There was one particularly low point after I graduated from my undergraduate and boomeranged back home to Burton-upon-Trent (a once prosperous brewery town), where I was working in a hospital folding the cancer appointment cancellation letters. It was cheaper than hiring a machine to do it. The office manager was my mum's friend; on reflection, one of the only middle-class connections my parents had and the furthest their influence could stretch. It was not a great job, but it was a job, and it helped reinforce the exact career direction I did not want to pursue. Eventually, I did land a job as a designer working at the now defunct Innovate 3, a small product design and technology develop-ment consultancy in the southeast of England, where I worked on a wide range of projects from Scalextric cars to single-use medical disposables (O'Callaghan, 2010). So, if I finally entered the design industry, why did I leap to academia? Why did I think, 'that is the career for me'?

In 2009, I was made redundant after what I now know as the Great Recession. The issue with small consultancies, or at least the one I worked for, is that a significant proportion of their income is generated from bigger companies' overspill and outsourcing of work. As the purse strings of the more prominent companies tighten, external expenditure is typically one of the first things to go into a recession. Work could no longer sustain the entire team, and few design companies had employment opportunities. I wanted job security, and academia at that time offered that (a short-term contract in a recession is at least a contract). I am not going to pretend I changed careers with a lofty ambition of wanting to save the planet through design, which sounds like the antithesis of the guidance this chapter is supposed to provide, but it was just the pragmatics of life for me at that stage; empathy and understanding of the negative costs of design was to come much later. When I was made redundant, and I was looking for design projects to keep me going, my partner-now-wife suggested I reach out to my old lecturers at Loughborough University to see if they had any design work going. I had worked freelance on a few design projects after my master's within the department (including refillable packing and behaviour change) (Lilley, 2007), so it seemed a promising direction. When I asked, they said no, but they did have a three-year PhD stipend if I was interested. I was not too fussed about what it was about at the time, as long as it involved design, but working with Senior Lecturer Dr Tracy Bhamra (currently Senior Vice-Prin-cipal Student & Staff Experience and Professor of Design for Sustainability at Royal Holloway, University of London) and Research Associate Dr Debra Lilley (currently Senior Lecturer in Design at Loughborough University) was undoubtedly the best – and least considered – decision of my working life.

The design researcher

Thankfully, it turned out that I immensely enjoyed, and was quite good at, the thing I was employed to do. I was a doctoral researcher (also known as a PhD student – which typifies the general confusion of whether the position is student or staff within the UK, a debate that will not fit within the pages of this chapter) on the EPSRC-funded Carbon, Control and Comfort (CCC) pro-ject (UKRI, 2009), which was investigating domestic energy reduction and control practices in social housing. It was fascinating to investigate the theory of designing for behaviour change and why people do what they do and how design can mediate and shape interactions (Lilley & Wilson, 2013; Wilson et al., 2015; Wilson et al., 2016), while working with other academics at all levels and from other institutions and disciplines. Working alongside Dr Emma Hinton, an expert in sustainable consumption and social practices, on data collection and analysis was particularly formative, as was working under the guidance of Prof Bhamra and Dr Lilley. The designing of interventions included a radiator feedback device prototype that would indicate whether the heating system was on or not, which could also change state depending upon whether the windows were open (Wilson, 2013). Also, I enjoyed the engagement with people, especially within the complex and proud mining community of Merthyr Tydfil, an area of South Wales from which I genetically herald. It felt like I had finally found a meaning and a purpose to design rather than just making 'stuff'. Designing consumer products will always be very satisfying in a visceral, primitive 'I made this' sense, but this felt more grounded and real and empowering as something to be proud of. Although I had always wanted to go back to the design industry, I had found a new direction in which to apply my family's "try your best" mantra and my own creative skills.

After the PhD, I then worked as a Research Associate, still at Loughborough University, on the EPSRC Low Effort Energy Demand Reduction (LEEDR) project (UKRI, 2010) with Dr Val Mitchell. This happened as a result of a colleague getting promoted and vacating the position—the most fortuitous of timing. The topic was not really what my PhD was in at all – the context was energy-related but had little to do with behaviour change (as I had studied it). However, I thought it was an exciting project and would bring interesting challenges. It was also a further contract for employment, and research associate jobs were, and increasingly are, hard to come by. I worked with engineers and social scientists and got quite into understanding their methods as well as exploring, applying, and combining their approaches within my practice. Indeed, one of the research concepts I explored was combining ethnographic narratives with energy data. Could we somehow add that quantitative edge to the qualitative understanding (or vice versa, depending upon your epistemological allegiance)? This eventually led to the People, Objectives, and Resources through a Time and Space framework (Wilson et al., 2014). On this project, I even learned about public engagement (Figure 5.1) and user experience (UX) design for the first time. Back then, I did not know what UX was, and I had always considered myself a designer of physical products. However, I was encouraged by Dr Mitchell to explore and upskill within that space, eventually bringing fresh ideas into the energy sector with concepts like Kairos and Future/Self.

The Kairos app concept was based on the insight that control of energy consumption is not always an option within everyday life, especially if living in a household with caring responsibilities. Kairos invited users to create new and bespoke activity interrelations driven by conditional logic. Networked

FIGURE 5.1 Universities Week 2014 at the Natural History Museum. Author on the right of photo

appliance activity could be digitally 'snapped' together to create new digital flows, for example, by remotely connecting a baby's cot (with accelerometer sensors) to a washing machine. If a baby is in light sleep, the washer's spin speed is automatically reduced, thereby reducing the noise created and allowing night-time washing, giving back control to tired parents. Future/Self was a concept predicated on the insight of trying to save time by 'bending' time. Here, you would enter the basic details of your life, such as age, education, job, family, and so forth, and then it finds somebody who has had the same life experience and is now, for example, ten years older than you. Time has been bent as you look into your future and ask the older and hopefully wiser 'you' those questions about whether you should invest in solar panels now or wait until your next home (Wilson et al., 2014).

I then returned to the EPSRC Closed Loop Emotionally Valuable E-waste Recovery (CLEVER) project (UKRI, 2013a) with Dr Debra Lilley. Again, this was facilitated by a series of events outside my control. Colleagues shuffled between posts, and the contract became available as my old one ended. This was my first time looking at material waste rather than energy waste (Wilson et al., 2015). The question I was particularly drawn to was why people keep technology, in this case mobile phones, for so long after they have finished using them. It is not quite connected to my previous body of academic outputs or thoughts, but again, it is an interesting question to seek to understand and work with colleagues I could learn from. Working with experts in life cycle analysis, we completed a study in 2015 in which we calculated that people typically use a phone as a communication device for one year and eleven months and go on to keep it in what we called 'hibernation' – the 'dead storage period' when it is stuffed in a drawer – for three years (Wilson et al., 2017). It is stored for longer than it is ever used. In other words, in the five years you own a phone, most of the time, it is not even being used, and that device could have been given to somebody else, preventing further devices from being manufactured. This has genuine implications for material resources. Indeed, I was interviewed by *The Guardian* (2017) about the findings, although paradoxically, I would wager this would not be a surprise – or news – to most people.

Around this time, I also started to work with Dr Lilley on the Creative Outreach for Resource Efficiency (CORE) project, an umbrella project designed to maximise outreach activities from CLEVER and other 'More of Less' EPSRC sandpit-funded projects (UKRI, 2013b). While strategically, one could easily argue the merit on paper of being involved with such a project – the impact was very much the new hot topic for funding bodies back then, and eventually, I was made a Co-Investigator of the project, which has not harmed my CV – I really cannot emphasise enough the value of getting involved with public engagement activities as an academic, and the earlier, the better. Some benefits are universal to public engagement: getting closer to the people most impacted by your research, building confidence in public speaking, and working out how to translate and communicate academic knowledge so that the audience understands and cares (or at least is not disinterested or anti-educated). Much like the tangible pleasure of hands-on design and making, the absolute pleasure of this type of working is in the engagement itself: seeing people you do not know genuinely interested in your work and allowing yourself to be open to new conversations and experiences. At the start of the project, I would not have imagined I would find myself talking about mobile phones and electronic waste with Key Stage Three girls in a secondary school (Madani Girls School in Leicester). I certainly would not have expected them to present their concept after a day of workshops for a new sustainable mobile phone in rap style, but this is one of my genuine career highlights. The resulting toolkit, co-developed with the CLEVER project team, has been quite successful. However, what will stay with me is that truly unexpected moment of spontaneity from students.

The future research leader

All the above undeniably helped me secure my design lectureship at Loughborough University, a combination of industry-leading design skills and a useful research portfolio. Nonetheless, I am not a prolific research writer and still struggle today with manuscript writing. However, anything I produce is of quality and, considering my folio beyond the traditional rotation of journal papers, conference papers, and book chapters to include web articles (Wilson & Wilson, 2016) and even poetry (Science Museum Group,

2022), has something interesting to say. An assumption at this point may be that the trajectory then accelerated. However, the reality was the opposite. It was a prolonged journey for an extended period without a destination. It would be fair to say that during my probation, research ambition was replaced with teaching ambition.

It took two events to bring me back to my research senses and hone my interest towards Net Zero. The first event was that I was excluded from my department's significant circular economy research proposal. It was not anything sinister and looking back I get the decision making, but it annoyed me enough to write a two-page article called 'My Perspective of Plastic Waste,' which I dumped onto Twitter (Wilson, 2018). The standout lines for me within the article compare material selection to Batman and Robin in that the "Dynamic Duo would always end up in some perilous scenario that we would have to tune in next week to see resolved. No one cared if Robin survived. Plastic is Robin." This form of creative writing with purpose was cathartic, and equally, the response was encouraging. As one example, my Twitter outburst led to interesting conversations with Dr Fredrik Henriksson in Sweden around 'Robin Materials', which inspired parts of his thesis (Henriksson, 2021). The second event was more structural. Prof Hua Dong took over leadership of the research group (at the time, called Sustainable Design Research Group, then changed to Responsible Design Research Group, the group I now lead) and Prof Roger Haslam took over my annual performance and development review. With both coming from another institution or another discipline, they did not see me as still the PhD student and were probing why, given my history, I was not pursuing my research agenda. Both made me pause for thought and gave me the confidence to move forward again, but this time with my intent and purpose. Prof Dong encouraged me to become the Co-Convenor of the Design Research Society Sustainability Special Interest Group (DRS SusSIG for short), and Prof Haslam challenged me to lead and submit my first significant research proposal, my first research proposal that was successfully funded and, incidentally, the largest UKRI NERC (Natural Environment Research Council) project grant award that Loughborough University has ever held (UKRI, 2023).

Today, I am currently the Principal Investigator of the Perpetual Plastic for Food to Go (PPFTG) project (UKRI, 2020): a three-year, £1.15 million, design-led research project to reduce the impact of food to go packaging. The PPFTG is a UKRI NERC project, funded by the Industrial Strategy Challenge Fund (ISCF) Smart Sustainable Plastic Packaging Challenge (SSPP), in which I get to work alongside excellent Loughborough University Co-Investigators Dr Fiona Hatton, Lecturer in Polymer Chemistry; Dr Rhodes Trimingham, Senior Lecturer in Design; Dr Elliot Woolley, Senior Lecturer in Sustainable Manufacturing; and Dr Nikki Clark, Lecturer in Design (whose PhD I co-supervised and was the genesis to the project proposal) (Clark, 2021; Clark et al., 2020). The project's overarching aim is to eliminate food to go single-use plastics through a design-led research approach; our contribution to the Net Zero challenge is through transitioning the food to go industry towards a circular economy of plastic packaging.

In starting the project, we initially looked at the consumer behaviour around packaging that one may consider simple. For example, how would someone dispose of a sandwich box with those ubiquitous triangular card packs everywhere in the UK? However, there were no straight answers. We were in the throes of COVID-19, and arguably attention was divided, but a third of those we asked indicated that they would recycle all the pack, a third general waste, and a third some (York et al., 2022). They had no consensus, with two-thirds not getting it right; this, for us, was an opportunity. In addition, we were also finding a lot of competing values within the supply chain, but anyone who works within the industry will not find that a surprise. When we talk about transitions to a circular economy in academia, sometimes friction and competition can become ignored, or overcoming it becomes a desk-based exercise. Walking around the factories and talking to the workers, I realised rapidly that identifying problems is moot unless we could provide tangible and workable business solutions to move forward with. Other aspects of the project include novel cleaning assurance methods to replace the industry standard of manual ATP swabbing (Nahar et al., 2022), and the investigation of photoluminescent nanoparticle inclusions (Larder & Hatton, 2022). Although I do not always understand every aspect of underlying science within the project, what I can understand is potential. This led to the development of an inventive method for a consumable-less track and trace system, for which we are currently applying for a patent.

I frequently get asked: 'Why is the PPFTG project led by *design*?' and not, I infer, from a traditional subject like engineering or one of the sciences. The simple reason is that design has its knowledge, methods, skills, and way of viewing the world that one cannot find, or get to a lesser extent, in other disciplines. For example, with the concept of 'frame creation', alternative perspectives of the problem might be introduced and challenge the conventional; it may become reframed and open new and potentially more radical and beneficial directions (Wilson et al., 2022). At one point in the project, I became obsessed with the idea that a scratch on packaging could be a new form of consumable-less label, an indicator, a signifier of information (quite often, the label on a pack is excluded from impact calculations, but it can be significant). Can we read scratches, scuffs, and dents and what would be the message? This led to a workshop with the 'Many Happy Returns' packaging project at the University of Sheffield. We explored new metaphors for plastic packaging (Wilson et al., 2023) and a working relationship in which we are starting to explore new project ideas together.

So, how am I going to end this chapter? Why am I providing you with this account of my research journey in such a candid way? Being a leader is not about crafting an unassailable persona about yourself that exudes excellence.

What I hope you take away is that, yes, sometimes you can be excellent at what you do, and that can generate opportunities, but sometimes it just does not. Not to be dismissive or cliché, but it can just be about being the right person in the right place at the right time, although you still need to be receptive to the opportunities when they present themselves, and then take them. This is my lived experience and how I started my career. So, if you think you are not getting ahead, it might be the circumstances and surroundings in which you find yourself. Yet, I would ask you to question those external factors before defaulting to the internal and questioning your capability or value. Having been dubbed a 'Future Research Leader' by others (Wilson, 2022), and if I can suppress my middle-class academic imposter syndrome for a moment, here is my key advice. Be curious. Be open to new ideas, new opportunities, and new collaborations. I believe there is significant personal and professional growth in being more open and flexible. By exposing yourself intellectually to different concepts and domains, you will develop and learn new skills and insights that add to your core. I started my career wanting to be a designer in the industry, yet here I am as a design academic. Without malleability and openness, I would not be where I am and have the conversations and impact I enjoy today. Also, sometimes just listening to or watching others from your discipline, or perhaps others not like you, such as engineers and social scientists (or even designers), will enrich your research perspective and practice. I still find immense joy and learning in listening to others, whether at a conference with like-minded researchers or, more recently, stretching my mind working with psychologists at the University of Sheffield. A PhD often encourages independent working on a specific topic, but this is the model for the qualification, for the training course for the qualification, and is not, in my opinion, the model for how all professional research practice should work. I will not pretend that I was born a great project leader, research group lead, or even a Net Zero 'Future Research Leader', but by being open and with a degree of 'bumbling' between opportunities as they present themselves and learning from each experience, taking those choices that my parents wished for me, I hope to enhance my practice, but also – and perhaps most importantly – enjoy the journey. I hope this chapter helps you, in some small way, to do so, too.

References

Berry, A. H., Collie, K., Acayo Laker, P., Noel, L.-A., Rittner, J., & Walters, K. (2022). *The Black Experience in Design: Identity, Expression & Reflection*. Allworth Press.

Clark, N. (2021). *The Importance of Understanding Consumer Behaviour in the Development of Food-to-go Packaging for a Circular Economy within the UK*. Loughborough University. https://doi.org/https://doi.org/10.26174/thesis.lboro.14074259.v1.

Clark, N., Trimingham, R., & Wilson, G. T. (2020). Incorporating Consumer Insights into the UK Food Packaging Supply Chain in the Transition to a Circular Economy. *Sustainability*, 12(15), 6106. https://doi.org/10.3390/su12156106.

The Guardian. (2017). The Gold, Silver and Precious Metals Hidden in Our Homes. *The Guardian.* https://www.theguardian.com/sustainable-electricals-with-wrap/2017/aug/30/recovering-critical-raw-materials-hidden-homes.

Henriksson, F. (2021). *On Material Selection and Its Consequences in Product Development.* Linköping University.

Larder, R., & Hatton, F. L. (2022). Enabling the Polymer Circular Economy: Innovations in Photoluminescent Labeling of Plastic Waste for Enhanced Sorting. *ACS Polymers Au*, 3(2), 182–201.

Lilley, D. (2007). *Designing for Behavioural Change: Reducing the Social Impacts of Product Use Through Design.* In Department of Design and Technology: Vol. Doctoral Thesis. Loughborough University.

Lilley, D., & Wilson, G. T. (2013). Integrating Ethics into Design for Sustainable Behaviour . *Journal of Design Research*, 11(3), 278–299. https://doi.org/10.1504/JDR.2013.056593.

Nahar, S., Sian, M., Larder, R., Hatton, F. L., & Woolley, E. (2022). Challenges Associated with Cleaning Plastic Food Packaging for Reuse. *Waste*, 1(1), 21–39.

O'Callaghan, T. (2010). A Breakthrough in Vaccine Preservation. *Time.* https://healthland.time.com/2010/02/17/a-breakthrough-in-vaccine-preservation/.

Oxford English Dictionary. (2023). bumblebee. *OED Online.* www.oed.com/view/Entry/24660.

Science Museum Group. (2022). *Celebrating National Poetry Day 2022.* https://www.sciencemuseumgroup.org.uk/blog/celebrating-national-poetry-day-2022/.

UKRI. (2009). *Carbon, Control and Comfort: User-centred Control Systems for Comfort, Carbon Saving and Energy Management.* https://gtr.ukri.org/projects?ref=EP%2FG000395%2F1.

UKRI. (2010). *LEEDR: Low Effort Energy Demand Reduction.* https://gtr.ukri.org/projects?ref=EP%2FI000267%2F1.

UKRI. (2013a). *CLEVER – Closed Loop Emotionally Valuable E-waste Recovery.*

UKRI. (2013b). *CORE: Creative Outreach for Resource Efficiency.* https://gtr.ukri.org/projects?ref=EP%2FK026429%2F1.

UKRI. (2020). *Perpetual Plastic for Food to Go (PPFTG).* https://gtr.ukri.org/projects?ref=NE%2FV01076X%2F1.

UKRI. (2023). *Grants on the Web.* http://gotw.nerc.ac.uk/department.asp?sb=vd&inst=363.

Wilson, C. D., & Wilson, G. T. (2016). Future Technology in the 'Star Trek' Reboots: Tethered and Performative. *PopMatters.* https://www.popmatters.com/future-technology-in-the-star-trek-reboots-tethered-and-performative-2495415057.html.

Wilson, G. T. (2013). *Design for Sustainable Behaviour: Feedback Interventions to Reduce Domestic Energy Consumption.* In Loughborough Design School: Vol. Doctor of Philosophy. Loughborough University.

Wilson, G. T. (2018). My Perspective of Plastic Waste. *Twitter.* https://twitter.com/designgarrath/status/1014471637356331008.

Wilson, G. T. (2022). *"I bomme, as a bombyll bee dothe": An (un)random research journey.* Net Zero Conference 2022. Research Journeys in/to Net Zero: Current and Future Research Leaders in the Midlands, UK.

Wilson, G. T., Bhamra, T. A., & Lilley, D. (2015). The Considerations and Limitations of Feedback as a Strategy for Behaviour Change. *International Journal of Sustainable Engineering*, 8(3). https://doi.org/10.1080/19397038.2015.1006299.

Wilson, G. T., Bhamra, T., & Lilley, D. (2016). Evaluating Feedback Interventions: A Design for Sustainable Behaviour Case Study. *International Journal of Design*, 10(2).

Wilson, G. T., Bridgens, B., Hobson, K., Lee, J., Lilley, D., Scott, J. L., & Suckling, J. R. (2015). Single Product, Multi-lifetime Components: Challenges for Product-Service System Development. In *PLATE 2015*.

Wilson, G. T., Clark, N., Hatton, F. L., Trimingham, R., & Woolley, E. (2022). Perspective Perpetual Plastic for Food to Go: A Design-led Approach to Polymer Research. *Polymer International*, 71, 1370–1375. https://doi.org/10.1002/pi.6401.

Wilson, G. T., Leder-Mackley, K., Mitchell, V., Bhamra, T., & Pink, S. (2014). *PORTS: An interdisciplinary and systemic approach to studying energy use in the home*. Ubi-Comp 2014 – Adjunct Proceedings of the 2014 ACM International Joint Conference on Pervasive and Ubiquitous Computing, 971–978. https://doi.org/http://doi.org/10.1145/2638728.2641551.

Wilson, G. T., Baird, H. M., Beswick-Parsons, R., Clark, N., Eman, S., Gavins, J., Greenwood, S., Lilley, D., Mattinson, P., Webb, T. L., Woolley, E., Woy, P., & York, N. (2023). *New metaphors for plastic packaging*. 5th PLATE 2023 Conference.

Wilson, G. T., Smalley, G., Suckling, J. R., Lilley, D., Lee, J., & Mawle, R. (2017). The Hibernating Mobile Phone: Dead Storage as a Barrier to Efficient Electronic Waste Recovery. *Waste Management*, 60, 521–533. https://doi.org/10.1016/j.wasman.2016.12.023.

York, N., Larder, R., Nahar, S., Wilson, G. T., Clark, N., Trimingham, R., Woolley, E., & Hatton, F. L. (2022). *Perpetual plastic for food to go: Consumer behaviour*. Materials Research Exchange (MRE) 2022 Conference.

6

SUSTAINABLE BY BEHAVIOUR

Hyunjae Daniel Shin

The love of human power

In 2003, after completing my mandatory military service, I was in agony trying to figure out what to do with my life. My first academic degree attempt in interior design failed in 1999, and I was on the verge of deciding whether I should continue pursuing a career in design. Sure, I had a strong interest in design because my only hobby back then was to hop on the bus and read design magazines until I arrived at the last bus stop, near my house. One day, I stumbled upon an article about Trevor Baylis' wind-up radio (Baylis, 1999). Designing for the other 90% drew me into the article. However, my decision to study 'design' was solely influenced by a friend who lived in a very sophisticated, modern house filled with designer furniture. I later realised that a design degree is not required to live in such a house. Nonetheless, the article inspired me enough to be determined that 'product design' is what I wanted to do, and six months later, I was on a plane to the United Kingdom to begin my second attempt at an academic degree in product design.

I have never let go of wanting to design something like a wind-up radio. What I loved about the wind-up radio was not necessarily its purpose, which was designed for those with no electricity access. I was fascinated by the interaction it creates with the users. Beyond the primary function of being a radio or one of those torches that use a similar dynamo mechanism, it offers an aesthetic interaction of playing with 'electricity'. I may be exaggerating the concept of human-powered products because I am a big fan of them; however, the actual value of a human-powered system is the 'free electricity' that didn't come with the product. Most human-powered products employ a well-engineered gear system that allows users to make human muscular exertion, which converts this energy into the product's functional power. The

DOI: 10.4324/9781003380566-9

required effort when using human power makes us feel visceral about what we take for granted from the grid. Perhaps at this point, I realised that design can also serve the purpose of 'raising awareness'.

Making a human-powered product

I was extremely fortunate to work for two different companies during my placement years. I have nearly two and a half years of placement experience working as a confectionary package designer for a large corporation in the United States and as an interior designer for a small design firm in Vietnam. Working for these two very different companies has allowed me to gain many practical design skills dealing with things like mass production, product road-mapping, and designing one-of-a-kind bespoke interior spaces and furniture for clients. I could have stayed in the industry longer to enjoy the life of earning more money, but it was time for me to return to school to create something I had always aspired to do – designing a human-powered product. During my time away from school, I spent significant time planning for my final-year project, which would be a human-powered product. What mattered was proposing new innovative human-powered products that could raise awareness of the benefits of using this sustainable and healthy energy source.

There are two main research approaches in sustainable development, each based on belief-driven values. An eco-centric perspective would argue that humankind's environmental problems can be put to rest only by forging a mutually beneficial partnership with nature. However, many with a tech-nocentric viewpoint contend that environmental concerns can be solved by investing more resources into technical innovation. A similar notion has been observed in the field of human-powered products. Allowing users to consciously exert their muscle energy in a mechanical system, such as a wind-up dynamo, is one popular method. The other method, parasitic harvesting (Starner, 1996), involves using the human body or its movement as a power source to power electronic devices, such as installing a dynamo onto a revolving door. Each method has its way of getting people to interact with electricity generation, either by making them do something on purpose or by using automatic systems that do not require people to change their behaviour. I made both kinds of human-power products for my final-year project. My initial objective for the project was to create a working prototype of a dynamo that would be used in two of my human-power concepts. It took me a long time to bother in-house electrical technicians at school to develop a hand-built dynamo. Figure 6.1 shows my final prototype of a pull-cord dynamo that can fully charge a 5.5-volt capacitor with three pulls and a fly-wheel added to generate more electricity even when muscle exertion has ceased. This dynamo system was then applied to two proposals: lighting with a pull-cord mechanism that lights up for five minutes when pulled three times, and a foldable walking stick with a dynamo wheel attached on the

FIGURE 6.1 Pull-cord generator prototype

end that produces electricity by rolling the wheel while you walk. I was pleased with the final grade for both of my projects, but I was even happier to be able to call myself a 'human-powered product designer'. Yet, the project left me with a further enquiry about how the concept of a human-powered product could be used. Will people use this as an everyday item? What value is there in recommending this as a sustainable product, and more importantly, will people love the idea of using human power as an everyday energy source for a sustainable purpose? However, then I realised I still needed to become a true expert.

Switching human power back on

I found myself writing an application for a research degree a few months after finishing my bachelor's degree. My final-year product led me to a funder who provided me with a stipend to begin a PhD study to find answers to my human-power enquiry. Of course, it began with a review of all human-powered products and literature. One of the significant paradigm shifts was the realisation that tools like the mop, pencil sharpener, and whisk have always been a part of daily life and have existed throughout history. However, since the invention of powerful motors that rely on electricity from the grid, humans have gradually used less of their muscle power. People may have developed prejudices against those electronics, believing that they are wiser, more efficient, and can persuade users that their needs are being met relatively quickly. Over-reliance on electricity was once an affordable choice; however, recent utility bill increases may have aided in changing our attitudes.

Furthermore, these electricity prices may still appear to be affordable, but electricity production impacts human health, for example through carbon emissions from burning coal. On the other hand, using human power can be viewed as a renewable energy source far less harmful to the environment, portable, available on demand, and relatively inexpensive. I could have decided to focus my research on improving dynamo output or designing for another innovative power generation interaction, but this left me with a critical research question: will people use this as an everyday product, and for what purpose?

My first task in searching for an answer was quantifying the benefit of using human-powered products. I claimed it is a sustainable and healthy resource, but I had no hard evidence. Jose Casamayor, a good friend who now works at the University of Sheffield, assisted me with this investigation by conducting a speculative experiment called Life Cycle Assessment (LCA). Arjen Jansen, who received the first PhD in human-powered products, conducted a similar experiment in which he compared a human-powered radio to a battery-consuming radio unit. His research concluded that the environmental impact of battery-powered radios would equal the value of human-powered radios in between 1.4 and 2.9 years; this is known as the 'break-even point' (Jansen, 2011, p.129). This is not to argue that the environmental benefit is always present when people use the device. However, as these devices no longer require batteries, reducing life-cycle environmental impact is a potential benefit (Jansen, 2011). Following Jansen's research, we compared the human-powered radio to the standard plug-in powered radio. Our findings show that it will take 9.6 years to break even on the environmental impact and 417 years to break even on the cost of ownership from saving electricity consumption from the grid.

The outcome of the LCA left me with another research question: how can design help people sustain the use of human-powered products in order to maximise this, to some extent, marginal benefit? As a result, the focus of the research has shifted to gaining knowledge about design strategies for behavioural change. Fortunately, scholars have produced a substantial amount of literature and prior work (Bhamra et al., 2011; Lilley, 2007; Lockton et al., 2009; Wever et al., 2008; Zachrisson & Boks, 2012). Each has a unique vocabulary for the behavioural intervention strategies they have introduced or tested, but all can be plotted along a dimension of 'power in decision making' (Wilson, 2013). The distribution of control between the user and the product distinguishes the strategies. One side of the axis employs a script to induce behavioural change, whereas power in behaviour change depends more on the user in action. The strategies on the other side of the axis use force or a penalty function to ensure a behavioural change or avoid unintended behaviour without requiring the user to make a conscious decision, which requires some ethical consideration. Norman (1999) gives an analogous terminology to these two distinct approaches: 'real' and 'perceived' affordance. Often, undermining forces on behavioural change can be eliminated by delegating all action to the device in question, also known as 'black-boxed script' (Jelsma, 1999). Again, this seemed similar to the distinction between technocentric and eco-centric approaches. Most of the prior works on a design for behavioural change did not conclude to suggest the most effective strategy. Nonetheless, a selection should account for the characteristics of the behaviour in question, such as the urgency of the behaviour problem or judging whether motivation or incentive would significantly impact on sustaining the changed behaviour. However, most of their work emphasises the notion that design can take a crucial role in inducing sustainable behaviour by purposefully shaping it.

Finding a single best strategy did not seem to attract me. However, I was more inclined to design a human-powered product that can be perceived as a viable solution to a sustainable lifestyle when used daily. It has been proposed that combining behavioural strategies is far more effective than applying the sum of the same strategies (Stern, 1999). So why not try designing such a combination – a design that can give a penalty for unsustainable behaviour and give motivation to sustain the positive behaviour? After reviewing additional articles related to sustaining changed behaviour, also known as 'internalisation', I built a prototype called the 'Whitebox'. The name was inspired by the term mentioned earlier, the 'black box script', which refers to a system with more power of behaviour than the user. As an act of defiance towards this notion, the prototype was named the 'Whitebox', which encourages users to make their choices, have control, and be empowered. The box had a very simple mechanism. It included a human-powered bicycle simulator and a battery-containing box with a four-digit display, which had an outlet for plugging in an electrical device like a TV or screen unit (Figure 6.2).

FIGURE 6.2 The 'Whitebox' system

When the user generates power by pedalling, the credit increases; when the user watches TV, the credit decreases, and when it reaches zero, the TV turns off. An hour of pedalling equalled two hours of TV viewing. The system facilitates a condition where the user must plan their day to find a time to pedal to watch TV. The 'Whitebox' system is then deployed to ten households interested in trying out the idea of powering their TV. Thankfully, 9 out of 10 households have adjusted well to the new behaviour, with one exception that never really got into it. It was a challenging study that required three visits to each household, collecting behavioural data, and capturing their attitudes through interviews. All participants shared their one-week experience of powering their TV, and what kind of motivation influenced their behaviour to be sustained. It was an intriguing study to document the process of making sense of using human power as a daily energy source (Shin & Bhamra, 2016). They mostly participated in the study because they recognised the importance of exercise and were concerned that most people were physically inactive. TV is one of the best home entertainment devices, but we all watch it from a sofa for some reason. People are indeed idle and sedentary while watching TV. For a very long time, no one has dared to challenge this conventional behaviour. If ten minutes of brisk exercise per day is enough to help combat high blood pressure, diabetes, weight issues, depression and anxiety, and musculoskeletal problems like lower back pain, encouraging such activity in exchange for access to TV and other electronic devices will significantly benefit users and prevent sedentary behaviour.

Cycling time varied in these nine homes. During the one-week experiment, the nine households burned 55,322 calories in total and travelled the equivalent of 1,218 km – a return road trip from London to Edinburgh. It was quite an accomplishment that most people appreciated. On the other hand, some of the results, such as the financial benefit, surprised people. The average saving per household for using the human-powered TV was £0.22. Most of them were very disappointed. However, it was discovered that most of participants had also become highly energy-sensitive. For example, they developed a new habit of turning off electronic devices when not in use. In a nutshell, I think people appreciated their empowerment from using an alternative energy consumption mode. Again, I would like to thank these households for participating in a weeklong effort to power their TV, which was meaningful to my research.

Nevertheless, I was curious how long it would take for such a system to break even, meaning that they gain actual benefit from either reducing carbon emissions or making other small financial gains. So, we repeated the LCA experiment on the 'Whitebox' system, and the results show that it would take four and a half years to become a product with less environmental impact, with economic payback beginning after 22 years of use (Shin et al., 2017). Indeed, none of these figures indicated a significant motivator. However, because the government spends millions of pounds battling diseases related to physical inactivity, this may provide some assistance to the government, and people may spend less money going to the gym. It was once a silly research idea, but it is very reassuring to see how it received much attention during the COVID-19 period when people were struggling to stay at home. More recently, the escalating energy prices have highlighted the benefit of using human power as an alternative and everyday energy source. I might not have come to these conclusions if I had chosen to focus more on the technocentric approach to human-powered products; instead, I let my research lead me in the direction of being more sceptical about the topic. I used to believe in human-powered products, but my scepticism enabled me to propose how they should be designed and used in the future. Changing perception is the most difficult challenge for any behaviour change study. I worked hard to show that while there are some short-term benefits to using human power (sustainable behaviour), other potential benefits could grow larger over time. For this reason, I titled my PhD thesis 'Switching the human-power back on' to convince others that trying is worthwhile. This PhD has suggested my next research path: how can 'design' change behaviour to achieve a more sustainable outcome, and how to sustain that behaviour through 'design'?

Design for sustainable behaviour

I had to let go of the research on human power as my research interests were leaning towards 'design for sustainable behaviour'. I am not sure if this was a coincidence, but my first lectureship started at Loughborough University, where I met with colleagues who had done remarkable work in this field. I did not spend much time at this institute, but I was delighted to work with the scholars whose work I had cited in my PhD thesis, and I felt so privileged to have the opportunity to strengthen my confidence that this is the field I want to pioneer. Since then, I have been officially telling others that my field of research is 'Design for Sustainable Behaviour'.

My first PhD student came from a social care background and was interested in design. I recommended to her that we almost replicate the study method I used for my PhD, but this time, we would focus on new user behaviour. We decided to tackle the children's sedentary behaviour (e.g., too much sitting) due to increased digital media in the domestic environment. She went further into how we can account for the internalisation process via design-led intervention. We carefully examined the function of 'feedback' and 'feedforward' in encouraging behaviour change. These strategies have been widely applied in energy conservation behaviour. Recently, energy suppliers have been designing smart metres to better support their decisions at home. They no longer simply give live information about current consumption, but this also allows people to project their behaviour pattern to set goals and modify their behaviour to meet their target. Often, many behavioural interventions provide limited feedback, which only gives the consequence of behaviour outcome. The energy bill is the best example where it is already too late to change the behaviour to reduce your consumption.

On the other hand, feedforward allows users to foresee the choices of action that are based on future scenarios or predictions of future states that can be speculative. Due to this notion, feedforward has much direct interconnection with the motivation of behaviour, especially when these changed behaviours are expected to be repeated. We then applied this aspect in the prototype, which we called a 'Knudgebox', which granted access to TV time for children when some physical activity was recorded. Similar to the 'Whitebox' mechanism, the 'Knudgebox' had an LED bar, indicating behavioural information: the bar length increased when the user pedalled the bike, and the bar length decreased when the user had the TV on. This time, we sent the prototypes out to 20 households. More sophisticated methods were used to capture their behaviour data and how their motivation changed during the intervention (Shin et al., 2022). Many questioned that the Whitebox experiment was too short to evaluate its effectiveness; therefore, we set the study period as a minimum of 90 days. We mainly put more effort into evaluating how our design generated self-regulated behaviour. We carefully examined how much empowerment they felt, which information type between feedback and feedforward was more effective in influencing their motivation, and whether they had acquired any autonomous motivation to continue carrying the intended behaviour. I later published an article suggesting these three dimensions into a framework of 'Internalisation of Sustainable Behaviour', which can help develop and select design strategies based on characteristics of targeted behaviour (Shin & Bull, 2019). These dimensions are often interrelated and interdependent, to induce a more efficient behaviour change, which may lead to a successful internalisation. Of course, further work was needed to apply this framework in more real-world cases to refine its dimensions and interdependency.

Design for sustainable urban living lab

A few years later, I received an offer from a Korean institution in Seoul (the capital city of South Korea). Yonsei University, where I currently work, is well known for its research excellence. It was the best decision for my research career because Korean institutions provide considerable academic research support. Upon arrival, I established my research lab and opened new modules related to my research area. However, coping with air pollution, rush hour, child education, finding time and a place to exercise, and other issues made living in the capital city difficult. Seoul is a densely populated city that accommodates roughly a quarter of the country's population. I anticipated that there would be numerous difficulties that design

could potentially address to assist citizens in living more sustainably. Therefore, I named my lab Design for Sustainable Urban Living Lab focusing on
Design for Sustainable Behaviour as a primary research topic. We now have
four ongoing projects focusing on bringing sustainability during the product,
service, and systems 'use-phase', which explore:

i tackling child sedentary behaviour through smart design intervention
  (other domains called this digital therapeutics);
ii developing a new navigation algorithm to help bike commuters have
  more expansive choices, such as suggesting healthier air routes or safer
  bike lanes;
iii designing a new smart meter for Korean homes; and
iv assessing life cycle impact of a human-powered hydroponic system.

More recently, I have opened a new module at a postgraduate level
called 'Design for Sustainable Behaviour'. In order to further refine my
three-dimensional framework, I have been enjoying working with students
to apply the framework to develop many case studies, such as tackling
recycling behaviour on campus, encouraging more home cooking to
replace food delivery services, encouraging family exercise at home, and
using VR to generate peer competition for increasing physical activity
among teenagers and reduce mobile use before sleep. These projects are
all contributing to advancing the design field for sustainable behaviour,
focusing on exploring ways to strengthen the design strategies to induce
more effective internalisation of sustainable behaviour.

Designerly way of researching (in the future)

As a researcher, I started to wonder whether there could be a better way to
conduct design research that might significantly impact society. As yet, there
are limited avenues for disseminating research works, such as journal publication, which requires consensus with reviewers on empirical evidence to
support our claims. It takes a lot of time and effort to meet the reviewers'
requests on the expected rigour of research work. I did not go down this
path. More than anything, my dynamo was meant to serve as an example of
what design 'can, could, would' do with the help of some as-yet-unproven
scientific principles.

My current PhD student, Eunsun, has been a tremendous asset in determining where to take my research. We have collaborated to launch a new
module that investigates how speculative design might be used to influence
future sustainable behaviour. We are looking into the potential of 'fun' and
'design fiction' as a vehicle to spark more debate, leading to more sustainable future choices for individuals. Speculative design is a step up from the
idea of feedforward in altering people's behaviour. We now speak with

various political comedians to contextualise how they employ the metaphor of 'gag' to elicit empathy and participation from their audience, which can help us understand how to use these techniques more effectively in the discipline of speculative design. These activities are much fun, and when we engage in sufficient curiosity and inquiry, we uncover numerous new areas for investigation.

One may argue that what researchers value themselves decides whether an approach to sustainable development is technocentric or eco-centric. Others may think technology will resolve the environmental dilemma, but people can be persuaded to change their attitudes, behaviours, and habits. Even though consumers pay more to buy an energy-efficient TV, they frequently leave them on when they are not watching them. Designing the use-phase of a product may matter more than designing a product to achieve sustainability. This is the value of the research field I like to pioneer. Of course, I will endeavour to construct more empirical studies to better quantify the benefits of engaging in sustainable behaviour, to investigate ways to generate ranges of effective design-led strategies, and to investigate future scenarios that can be more influential in raising awareness about how we should behave differently in order to better shape the sustainable future.

My research activities may not appear to be directly related to net zero emission research, but I began a research journey by attempting to create a product that runs on renewable energy from human muscle, which I later discovered provides no meaningful benefit to the environment unless people properly use it. Then, I considered how we might encourage individuals to utilise human-powered devices more competently to achieve more significant outcomes and develop case studies to reinforce the empirical discoveries of my research. Beyond merely lowering carbon emissions, these outcomes provided options and empowerment for a well-being lifestyle, such as improving health or lowering risks. Now, I am at a crossroads, contemplating how things could be designed to help people behave more sustainably in the future. In the same way, as I consider my next move, here are my remaining questions for further research: How can we properly assess the effectiveness of design that influences behaviour for long-term sustainability? How can we better speculate a future that will encourage individuals to engage in more sustainable behaviour?

References

Baylis, T. (1999). Spring operated current generator for supplying controlled electric current to a load (United States Patent No. 5917310). http://www.freepatentson line.com/5917310.html.
Bhamra, T., Lilley, D., & Tang, T. (2011). Design for Sustainable Behaviour: Using Products to Change Consumer Behaviour. *The Design Journal*, 14 (4), 427–445. https://doi.org/10.2752/175630611X13091688930453.

Jansen, A. (2011). *Human Power Empirically Explored* [PhD Thesis, Delft University of Technology]. Delft, Netherlands.

Jelsma, J. (1999). Philosophy Meets Design, or How the Masses Are Missed (and Revealed again) in Environmental Policy and Eco Design. *Consumption, Everyday Life and Sustainability*, Reader for ESF Summer School 1999, Lancaster University.

Lilley, D. (2007). *Designing for Behavioural Change: Reducing the Social Impacts of Product Use through Design* [PhD Thesis, Loughborough University]. Loughborough, UK.

Lockton, D., Harrison, D., Holley, T., & Stanton, N. A. (2009). *Influencing interaction: Development of the design with intent method.* Proceedings of the 4th International Conference on Persuasive Technology.

Norman, D. A. (1999). Affordance, Conventions, and Design. *Interactions*, 6 (3), 38–43. https://dl.acm.org/doi/pdf/10.1145/301153.301168.

Shin, H. D., Al-Habaibeh, A., & Casamayor, J. L. (2017). Using Human-powered Products for Sustainability and Health: Benefits, Challenges, and Opportunities. *Journal of Cleaner Production*, 168, 575–583. https://doi.org/10.1016/J.JCLEPRO.2017.09.081.

Shin, H. D., & Bhamra, T. (2016). Design for Sustainable Behaviour: A Case Study of Using Human-power as an Everyday Energy Source. *Journal of Design Research*, 14 (3), 280–299. https://doi.org/10.1504/jdr.2016.079763.

Shin, H. D., & Bull, R. (2019). Three Dimensions of Design for Sustainable Behaviour. *Sustainability*, 11 (17), 4610. https://www.mdpi.com/2071-1050/11/17/4610/pdf.

Shin, H. D., Nwankwo, F., & Al-Habaibeh, A. (2022). Design-led Intervention to Reduce Sedentary Behavior in Young People. *She Ji: The Journal of Design, Economics, and Innovation*, 8 (3), 387–414. https://doi.org/https://doi.org/10.1016/j.sheji.2022.10.003.

Starner, T. (1996). Human-powered Wearable Computing. *IBM Systems Journal*, 35 (3–4), 618–629. doi:10.1147/sj.353.0618.

Stern, P. C. (1999). Information, Incentives, and Proenvironmental Consumer Behavior. *Journal of Consumer Policy*, 22 (4), 461–478. https://link.springer.com/content/pdf/10.1023/A:1006211709570.pdf.

Wever, R., van Kuijk, J., & Boks, C. (2008). User-centred Design for Sustainable Behaviour. *International Journal of Sustainable Engineering*, 1 (1), 9–20. https://doi.org/10.1080/19397030802166205.

Wilson, G. T. (2013). *Design for Sustainable Behaviour: Feedback Interventions to Reduce Domestic Energy Consumption*. In Loughborough Design School: Vol. Doctor of Philosophy. Loughborough University.

Zachrisson, J., & Boks, C. (2012). Exploring Behavioural Psychology to Support Design for Sustainable Behaviour Research. *Journal of Design Research*, 10 (1), 50–66. https://www.inderscienceonline.com/doi/abs/10.1504/JDR.2012.046139.

7

SUSTAINABLE PRODUCTION AND CONSUMPTION BY UPCYCLING TOWARDS NET ZERO

Kyungeun Sung

The start of upcycling research

I am originally from South Korea. I did my first degree (BSc) in Industrial Design at Korea Advanced Institute of Science and Technology, then a master's degree (MSc) in Strategic Product Design at Delft University of Technology in the Netherlands. I came to the UK in September 2013 to embark on my PhD journey on 'Sustainable production and consumption by upcycling: Understanding and scaling up niche environmentally significant behaviour' (Sung, 2017) in the School of Architecture, Design and the Built Environment at Nottingham Trent University (NTU). My PhD research project was funded by the Vice Chancellor's PhD scholarship from NTU with support from the Centre for Industrial Energy, Materials and Products (CIE-MAP), funded by the EPSRC (Engineering and Physical Sciences Research Council), grant number EP/N022645/1. As the aim of the CIE-MAP was to identify all the opportunities along the product supply chain that ultimately deliver a reduction in industrial energy use, the starting point of my PhD was looking at different design strategies to reduce industrial energy consumption. Amongst different strategies I explored upcycling as one promising opportunity to contribute to industrial energy use reduction, informed and inspired by my supervisors, fellow students, and colleagues in CIE-MAP.

For those who are not familiar with the term upcycling, it is an umbrella concept incorporating a variety of design and/or material processes based on used or waste products, components, and materials in order to create a new or modified product, artefact, or material of higher quality or value than the compositional elements. Examples of upcycling are 'creative' or 'innovative' repair, refurbishment, redesign, upgrade, and remanufacture, all of which are also an essential part of the circular economy – a sustainable alternative to the

DOI: 10.4324/9781003380566-10

traditional linear economy of take, make, use, and dispose (Stahel, 2016). Upcycling, in principle, decreases solid wastes, increases resource/material efficiency, and reduces industrial energy consumption by manufacturing with virgin materials, lowering greenhouse gas emissions (ultimately towards net zero).

There are numerous ways that I could have explored upcycling, but my literature review at the time showed the current state of knowledge, gaps in knowledge, and a way forward (Sung, 2015b). Accordingly, I decided to focus on consumer/citizen upcycling at the household and community level to fill the knowledge gap. The very first paper I wrote and presented was a theoretical literature review to explore the possible determinants of individual upcycling based on a social psychological framework (Sung, Cooper, & Kettley, 2014), which was further developed into my main data collection for my PhD. I was also interested in how upcycling affects the emotional bond between users/upcyclers and products (or product attachment), and, therefore, the extended use of upcycled products. Hence, I conducted a short questionnaire study to understand the potential links between upcycling, product attachment, and product longevity (Sung, Cooper & Kettley, 2015a; 2015b).

In the second year of my PhD, I was introduced to Sarah Turner, the Nottingham-based artist and designer through upcycling craft (who also studied at NTU). I interviewed her and wrote a portrait paper (Sung & Cooper, 2015). Despite its relatively low value as a research paper, this portrait paper was significant to me as my first journal article publication, through which I experienced and learned about the painful process of academic publication. Another significant moment was when my second supervisor, Prof Sarah Kettley, invited me to contribute to the Bloomsbury Encyclopaedia of Design. So I wrote short sections on appropriate technology, renewable resources, source reduction, and waste (Sung, 2015a). It was the first time I published anything in the form of a book. In 2016, as I approached the end of my PhD, I attended the Design Research Society Conference. I presented the overarching methodological framework of my PhD (Sung, Cooper, & Kettley, 2016). When I finished the draft of my PhD thesis after cutting out the side projects and focusing on one coherent story, it turned out to be really about how and why UK makers/citizens upcycle items at home (or at the community level), and what design and policy interventions could be developed and implemented to scale this up in households and beyond in the UK and elsewhere.

Continuation of upcycling research after PhD as Research Associate and Lecturer

With the PhD thesis submission, I was fortunate to start a part-time Research Associate (RA) position at the same school in January 2017, funded by the NTU Materials Research Seed-Corn Fund, while waiting for my viva. The postdoctoral position was to explore challenges and opportunities for scaling up small and medium-sized enterprises (SMEs) based on upcycling in the UK. I worked with my first supervisor, Prof Tim Cooper, Dr Jagdeep Singh, and other colleagues from Nottingham Business School. This work was the extended version of our initial research (one of my PhD side projects) into the challenges and support for scaling up upcycling businesses in the UK based on the workshop for small upcycling business owners (Sung, Cooper, Ramanathan, & Singh, 2017). The main RA work was interviewing upcycling businesses in the fashion industry in the UK (Sung et al., 2017) and UK furniture upcycling businesses. In addition to this RA work, I also had other tasks such as the PhD viva, minor revisions to the PhD thesis, writing papers from the PhD thesis, continuing collaborative work as part of CIE-MAP, and applying for a permanent position. The first decent publication to come out of my PhD was a book chapter based on my interview study asking UK citizens with practical upcycling experiences about how they upcycle items and why they do so (Sung, Cooper, & Kettley, 2017). One of the CIE-MAP work package reports I contributed to ('Understanding consumption: Why and how we use products') was published four years after the project (Salvia et al., 2017). After graduating in July 2017, I was again very fortunate to be offered a full-time, permanent position as a Lecturer in Product Design at De Montfort University (DMU) in Leicester, UK, starting in September 2017.

The first year of teaching (or teaching and research) was tough, probably more challenging than I imagined. There was a probationary period of one year, so I was nervous and worried about the uncertainty. Everything was new to me regarding the environment, the system, the people, and some of the tasks I had to do. It was a steep learning curve. With a bit of struggle, I continued the NTU RA work and presented our study on British fashion upcycling businesses at the Global Fashion Conference (Sung, Cooper, Oehlmann, & Singh, 2018). Based on my PhD, I also published a book chapter on understanding and scaling up upcycling in the UK from the perspectives of emerging social movements for sustainability (Sung, Cooper, & Kettley, 2018). My good old friend from the University of Liverpool, Dr JungKyoon Yoon (now at Cornell University), invited me to deliver a student design workshop on 'Sustainable design and product development with the circular economy and upcycling' for the second-year Industrial Design students, which was a good change and something new to do. Then, we presented the workshop results framed as how we embed sustainability in design education (Sung & Yoon, 2019). I worked with the sustainable artist and sculptor, Michelle Reader, to organise my first public engagement event, 'Art with Upcycling' (art and craft co-creation workshop based on upcycling for families with young children), in Leicester Museum & Art Gallery as part of the Being Human 2018 Festival.

In 2019, after the probation year and getting used to the new workplace and workloads, I could focus on re-writing my PhD and RA work to publish them as journal articles. The first journal article published in 2019 came from my PhD online survey on the factors influencing upcycling for UK makers (Sung, Cooper, & Kettley, 2019b). The second published article was about my PhD – developing policy and design interventions for scaling up UK upcycling in households and beyond based on interviews, surveys, and semi-Delphi studies (Sung, Cooper, & Kettley, 2019a). The third published article was on the challenges and opportunities for scaling up textile and wood upcycling businesses in the UK, based on my work at NTU as an RA (Singh, Sung, Cooper, West, & Mont, 2019).

I had some spare time to do a quick and dirty literature review on upcycling for teaching and learning in higher education and published a book chapter from it (Sung, 2019). Also in 2019 I received an endorsement from the Royal Academy of Engineering for exceptional promise as a potential world leader in upcycling research. In 2020, the last journal article from the NTU RA work was published on multi-stakeholder perspectives on scaling up UK fashion upcycling businesses (Sung, Cooper, Oehlmann, Singh, & Mont, 2020). With the pandemic and the lockdown, 2020 was a difficult and challenging year. However, we (Dr Ben Bridgens, Dr Jagdeep Singh, Prof Tim Cooper, and myself) managed to organise the 'International Upcycling Symposium 2020: Research and Practice' as a joint online event between DMU, Lund University in Sweden, NTU, and Newcastle University. The symposium was a reasonable success, with excellent presentations and a good turnout.

Development of upcycling research as a Senior Lecturer

In 2021, I became a Senior Lecturer. I thought it was not a big deal. Everyone becomes a Senior Lecturer at some point in any of the higher education institutions in the UK. But certainly, my roles have changed or evolved. I became a Programme Leader for MA and MSc Product Design, the School's (Art, Design and Architecture) Lead Ethics Reviewer, and Faculty (Arts, Design and Humanities) Deputy Head of Research Ethics. I was also selected for a year-long DMU Vice-Chancellor's Future Research Leaders Programme. I welcomed all the new challenges and opportunities. It became another busy year with teaching and extra research activities. I conducted and presented two case studies evaluating interventions for scaling up upcycling: one was based on a community event in the UK, and another on the upcycling plaza in South Korea (Sung, 2021a). I wrote a foreword for Chris Brosse's book, *Garbage Does Not Exist: Towards Upcycling and the Circular Economy* (*La Basura No Existe: Hacia el suprarreciclaje y la Economía Circular*) (Sung, 2021b).

Based on the International Upcycling Symposium 2020, we submitted a book proposal to Springer and worked on the edited book publication (Sung, Singh, & Bridgens, 2021). For this edited book, I was involved in writing several chapters either by myself or with my colleagues and collaborators, such as 'Introduction: State-of-the-art upcycling research and practice' (Sung, Singh, Bridgens, & Cooper, 2021), 'Upcycling of silicon solar cells: What are the options?' (Isherwood & Sung, 2021), 'Systems approach to scaling-up global upcycling: Framework for empirical research' (Singh & Sung, 2021), 'Repair and upcycling: How do we know which repair is considered as upcycling?' (Sung & Dao, 2021), 'Understanding quality in upcycled products' (Sung & Mahajan, 2021), 'Understanding and measuring value and quality of upcycling with fuzzy linguistic approach' (Sung, 2021c), and 'Designing for positive upcycling experiences with people's well-being in mind' (Sung & Yoon, 2021). The positive side effect of the very unfortunate pandemic and lockdown was that it increased my academic writing productivity, and I published my first-ever edited book (Sung, Singh, & Bridgens, 2021).

I also wrote another book chapter based on a literature review to identify challenges and opportunities for scaling up global upcycling towards sustainable production and consumption (Sung & Abuzeinab, 2021). Furthermore, with my increased productivity, I developed and applied for an AHRC (Arts and Humanities Research Council) Research Networking grant. As well as the academic writing, over the summer, I organised an 'Upcycling Art, Craft and Design Competition' (open to DMU students, alumni, staff, and people in the Leicester community), funded by the DMU Sustainability Team and in partnership with LCB (Leicester Creative Businesses) Depot. In October, we (James Burkmar from LCB Depot and myself) also used the competition entries to create an 'Upcycling Art, Craft and Design Exhibition' as part of LCB Depot's annual Design Season.

In 2022, I received the great news that I had been awarded the AHRC Research Networking grant for the International Upcycling Research Network (myself as PI and Prof Richie Moalosi from the University of Botswana as Co-I). This was the first proper external research funding I received after three unsuccessful attempts to secure British Academy/Leverhulme Small Research Grants between 2018 and 2020. I also took on other roles: I became a member of the AHRC Peer Review College, a reviewer on the Institute of Art and Design Research Funding Committee, and a Faculty Early Career Representative on the Concordat for Researcher Development Task and Finish Group. As the lead guest editor, I edited the special issue of the *Sustainability* journal on 'Sustainable consumption and production by upcycling: Advances in science and practices'. The special issue had four excellent papers, including my paper on adapting Darnton's nine principles framework for behaviour change with the UK upcycling case study (Sung, Cooper, & Kettley, 2022).

The AHRC-funded International Upcycling Research Network project started in June 2022. We currently have over 60 network members from 19 countries across five continents contributing to the network's activities and outcomes. The first network event we organised was the Net Zero Conference 2022 at DMU with Dr Patrick Isherwood from Loughborough University on 24th June 2022. In September 2022, we used an Interpretive Structural Modelling workshop (Abuzeinab, Arif, & Qadri, 2017) facilitated by Dr Amal Abuzeinab to collaboratively investigate and identify critical global challenges and opportunities for scaling up upcycling businesses. In September, we also hosted Upcycling Station, a drop-in event as part of the LCB Depot Takeover of the British Science Festival 2022 in partnership with DMU, where we screened some informative and inspirational videos made by the network members and offered a hands-on participatory upcycling workshop facilitated by Dr Mary O'Neill and her Fine Art students at DMU. From September 2022 to January 2023, we expanded our understanding of current upcycling research and practice across industries, disciplines, and countries by hosting seminars with global experts: Upcycling in Africa (September), Asia (October), the Americas (November), and Australia and Europe (January). The first collaborative paper that emerged from the network project was an exploratory multiple case study of successful upcycling businesses (Sung, Hughes, & Hsu, 2022). The second collaborative paper was on the network project – promoting upcycling through an international research network (Moalosi & Sung, 2022). At the end of 2022, I received the DMU Climate Change Awards 2022 Climate Action in Arts and Culture, based on all my work on upcycling for net zero (the list of activities and publications above).

At the beginning of 2023, I had the pleasure of publishing another collaborative research as a journal article – predictors of upcycling in the highly industrialised West from the online survey study conducted in three continents of Australia, Europe, and North America (Sung, Ku, & Yoon, 2022). In February 2023, I organised a UK upcycling networking event at DMU with the UK International Upcycling Research Network members, DMU staff and students working on other upcycling-related projects, and external people who are interested in the network. In June 2023, I presented two papers from the network project at the Product Lifetimes and the Environment (PLATE) 2023 conference. One paper looks at key global challenges and opportunities for scaling up upcycling businesses, based on the preliminary analysis of the Interpretive Structural Modelling workshops (Abuzeinab et al., 2023). Another paper looks at how to understand and teach upcycling in the context of the circular economy, based on a literature review and the first phase of the Delphi with global experts in upcycling and the circular economy (Sung et al., 2023). In July 2023, I presented two other papers at ICED (International Conference on Engineering Design) 2023, published as conference proceedings with Cambridge University Press. One is a review paper on understanding upcycling and the circular economy and their interrelationships for design education (Sung, 2023). Another paper is based on the major project of our excellent DMU MSc Product Design student, Alex Heaton, on accessible solar energy technology for domestic use in the UK (Heaton, Sung, & Isherwood, 2023).

There were two short-term projects until the end of July 2023. One project was 'New product development for upcycling and the circular economy', which aimed to develop radical innovations for everyday multi-component and mixed material electronics (e.g., kettles, vacuum cleaners) so that they can be upcycled and comply with circular economy principles, funded by DMU HEIF (Higher Education Innovation Funding). Another project was 'Renewable energy system design innovation for the circular economy', a scoping study based primarily on a literature review to identify the knowledge gap in this area, funded by DMU Net Zero and Climate Action Funding. I am preparing for the International Upcycling Festival 2024 as the culminating event of the AHRC-funded International Upcycling Research Network. We (Prof Jinsong Shen and myself) appointed a new PhD student on a DMU Doctoral College Scholarship for a project, 'Fashion and textile upcycling for new product development', starting in January 2024. For the rest of 2023 (it is September now), I will be working on two journal articles, building on the conference papers on the key global challenges and opportunities for scaling up upcycling businesses in the world, and how to understand and teach upcycling in the context of the circular economy.

Future research in upcycling and other related areas

In 2024, I will focus on successfully organising the International Upcycling Festival 2024 by April and submitting the final report to the funder, AHRC, by May. I will use the rest of the year to work on the applications for the next external funding. I have three ideas so far. One is to focus on the top two or three key challenges and opportunities for scaling up upcycling worldwide and conduct in-depth studies on each factor, including action research (i.e., piloting promising interventions to address the key challenges and monitoring the results). Another idea is to scale up the 'New product development for upcycling and the circular economy' project nationally, involving several disciplines/sectors with relevant industry partners. The third idea is a research and design project based on the scoping research results from the 'Renewable energy system design innovation for the circular economy' project. With my future research projects, I hope to make a real difference and impact by scaling up upcycling in homes and industries in the UK and beyond. I want to explore various scalable upcycling ideas, concepts, processes, and techniques across different industries/sectors to identify and promote best practices. In particular, I believe upcycling in the renewable energy sector (i.e., creating a circular renewable energy sector) is paramount to achieving net zero.

References

Abuzeinab, A., Arif, M., & Qadri, M. A. (2017). Barriers to MNEs green business models in the UK construction sector: An ISM analysis. *Journal of Cleaner Production*, 160, 27–37.

Abuzeinab, A., Sung, K., Moalosi, R., Satheesan, A., Garba, B., Adeh, F., Lim, H., Baek, J., & Njeru, S. (2023). *Key global challenges and opportunities for scaling up upcycling businesses in the world: Interpretive structural modelling workshop preliminary analysis*. Product Lifetimes and the Environment (PLATE) 2023 Conference, Espoo, 31 May–2 June.

Heaton, A., Sung, K., & Isherwood, P. (2023). *Accessible solar energy technology for domestic applications in the UK: Edge Solar*. Proceedings of the International Conference on Engineering Design (ICED23), Bordeaux, France, 24–28 July.

Isherwood, P., & Sung, K. (2021). Upcycling of silicon solar cells: What are the options? In K. Sung, J. Sing, & B. Bridgens (Eds.), *State-of-the-Art Upcycling Research and Practice: Proceedings of the International Upcycling Symposium 2020* (pp. 19–24). Springer.

Moalosi, R., & Sung, K. (2022). *Promoting upcycling through an international research network*. International Online Conference on Reuse, Recycling, Upcycling, Sustainable Waste Management and Circular Economy (ICRSC) 2022. Online. 9–11 September.

Salvia, G., Braithwaite, N., Moreno, M., Norman, J., Scott, K., Sung, K., Cooper, T., Barret, J., & Hammond, G. (2017). Understanding consumption: Why and how do we use products? Centre for Industrial Energy, Materials and Products (CIE-MAP). https://dora.dmu.ac.uk/handle/2086/14780.

Singh, J., & Sung, K. (2021). Systems approach to scaling-up global upcycling: Framework for empirical research. In K. Sung, J. Sing, & B. Bridgens (Eds.), *State-of-the-Art Upcycling Research and Practice: Proceedings of the International Upcycling Symposium 2020* (pp. 99–103). Springer.

Singh, J., Sung, K., Cooper, T., West, K., & Mont, O. (2019). Challenges and opportunities for scaling up upcycling businesses – the case of textile and wood upcycling businesses in the UK. *Resources, Conservation and Recycling*, 150, 104439.

Stahel, W. R. (2016). The circular economy. *Nature*, 531(7595), 435–438.

Sung, K. (2015a). Appropriate technology; renewable resource; source reduction; waste. In C. Edwards (Ed.), *Bloomsbury Encyclopedia of Design*. Bloomsbury Academic.

Sung, K. (2015b). A review on upcycling: Current body of literature, knowledge gaps and a way forward. *Proceedings of the 17th International Conference on Environmental, Cultural, Economic and Social Sustainability*, 17(4), 28–40.

Sung, K. (2017). *Sustainable production and consumption by upcycling: Understanding and scaling up niche environmentally significant behaviour* [PhD thesis, Nottingham Trent University]. NTU Institutional Repository. https://irep.ntu.ac.uk/id/eprint/31125/.

Sung, K. (2019). Upcycling for teaching and learning in higher education: Literature review. In W. L. Filho, A. L. Salvia, R. W. Pretorius, L. L. Brandli, E. Manolas, F. Alves, U. Azeiteiro, J. Rogers, C. Shiel, & A. D. Paco (Eds.), *Universities as Living Labs for Sustainable Development: Supporting the Implementation of the Sustainable Development Goals* (pp. 371–382). Springer.

Sung, K. (2021a). *Evaluating two interventions for scaling up upcycling: Community event and upcycling plaza* [conference paper]. International Conference on Resource Sustainability (icRS) 2021, Dublin, Ireland.

Sung, K. (2021b). Prólogo (foreword). In C. Brosse (Ed.), *La Basura No Existe: Hacia el suprarreciclaje y la Economía Circular (Garbage Does Not Exist: Towards Upcycling and the Circular Economy)* (pp. 14–16). Edinexo.

Sung, K. (2021c). Understanding and measuring value and quality of upcycling with fuzzy linguistic approach. In K. Sung, J. Sing, & B. Bridgens (Eds.), *State-of-the-Art Upcycling Research and Practice: Proceedings of the International Upcycling Symposium 2020* (pp. 127–130). Springer.

Sung, K. (2023). *Understanding upcycling and circular economy and their interrelationships through literature review for design education.* In: Proceedings of the International Conference on Engineering Design (ICED23), Bordeaux, France, 24–28 July.

Sung, K., & Abuzeinab, A. (2021). Challenges and opportunities for scaling up global upcycling towards sustainable production and consumption: Literature review. In W. L. Filho, A. Azul, F. Doni, & A. Salvia (Eds.), *Handbook of Sustainability Science in the Future: Policies, Technologies and Education by 2050* (pp. 1–19). Springer.

Sung, K., & Cooper, T. (2015). Sarah Turner – Eco-artist and designer through craft-based upcycling. *Craft Research*, 6(1), 113–122.

Sung, K., Cooper, T., & Kettley, S. (2014). *Individual upcycling practice: Exploring the possible determinants of upcycling based on a literature review* [conference paper]. Sustainable Innovation 2014 Conference, Copenhagen, Denmark.

Sung, K., Cooper, T., & Kettley, S. (2015a). *An exploratory study on the consequences of individual upcycling: Is it worth making people feel attached to their upcycled products?* [conference paper]. NTU ADBE Doctoral Conference, Nottingham, UK.

Sung, K., Cooper, T., & Kettley, S. (2015b). An exploratory study on the links between individual upcycling, product attachment and product longevity. *Proceedings of the Product Lifetimes and the Environment 2015 Conference*, 349–356.

Sung, K., Cooper, T., & Kettley, S. (2016). An alternative approach to influencing behaviour: Adapting Darnton's nine principles framework for scaling up individual upcycling. *Proceedings of DRS2016: Design + Research + Society – Future-Focused Thinking*, 1–14.

Sung, K., Cooper, T., & Kettley, S. (2017). Individual upcycling in the UK: Insights for scaling up towards sustainable development. In W. L. Filho (Ed.), *Sustainable Development Research at Universities in the United Kingdom* (pp. 193–227). Springer.

Sung, K., Cooper, T., & Kettley, S. (2018). Emerging social movements for sustainability: Understanding and scaling up upcycling in the UK. In R. Brinkmann, & S. Garren (Eds.), *The Palgrave Handbook of Sustainability* (pp. 299–312). Palgrave Macmillan.

Sung, K., Cooper, T., & Kettley, S. (2019a). Developing interventions for scaling up UK upcycling. *Energies*, 12(14), 2778.

Sung, K., Cooper, T., & Kettley, S. (2019b). Factors influencing upcycling for UK makers. *Sustainability*, 11(3), 870.

Sung, K., Cooper, T., & Kettley, S. (2022). Adapting Darnton's nine principles framework for behaviour change: The UK upcycling case study. *Sustainability*, 14(3), 1919.

Sung, K., Cooper, T., Oehlmann, J., & Singh, J. (2018). *Scaling up British fashion upcycling businesses: Multi-stakeholder perspectives* [conference abstract]. Global Fashion Conference, London, UK.

Sung, K., Cooper, T., Oehlmann, J., Singh, J., & Mont, O. (2020). Multi-stakeholder perspectives on scaling up UK fashion upcycling businesses. *Fashion Practice*, 12(3), 331–350.

Sung, K., Cooper, T., Painter-Morland, M., Oxborrow, L., Ramanathan, U., & Singh, J. (2017). *Multi-stakeholder perspectives on the challenges and success factors for scaling up upcycling businesses in fashion industry in the UK* [conference abstract]. 18th European Roundtable for Sustainable Consumption and Production, Skiathos, Greece.

Sung, K., Cooper, T., Ramanathan, U., & Singh, J. (2017). Challenges and support for scaling up upcycling businesses in the UK: Insights from small-business entrepreneurs. In C. Bakker, & R. Mugge (Eds.), *Proceedings of the Product Lifetimes and the Environment (PLATE) 2017 Conference* (pp. 397–401). IOS Press.

Sung, K., & Dao, T. (2021). Repair and upcycling: How do we know which repair is considered as upcycling? In K. Sung, J. Sing, & B. Bridgens (Eds.), *State-of-the-Art Upcycling Research and Practice: Proceedings of the International Upcycling Symposium 2020* (pp. 105–109). Springer.

Sung, K., Hughes, P., & Hsu, J. (2022). *Exploratory multiple case study on successful upcycling businesses: ChopValue, Freitag and Pentatonic* [conference paper]. British Academy of Management (BAM) 2022 Conference, Manchester, UK.

Sung, K., Ku, L., & Yoon, J. (2022). Predictors of upcycling in the highly-industrialised west: A survey across three continents of Australia, Europe, and North America. *Sustainability*, 15(2), 1461.

Sung, K., & Mahajan, D. (2021). Understanding quality in upcycled products. In K. Sung, J. Sing, & B. Bridgens (Eds.), *State-of-the-Art Upcycling Research and Practice: Proceedings of the International Upcycling Symposium 2020* (pp. 119–122). Springer.

Sung, K., Moalosi, R., Satheesan, A., Brosse, C., Burton, E., Lim, H., Cheung, K., Debrah, R., Lettmann, S., & Gaukrodger-Cowan, S. (2023) *How to understand and teach upcycling in the context of the circular economy: Literature review and first phase of Delphi*. Product Lifetimes and the Environment (PLATE) 2023 Conference, Espoo, 31 May–2 June.

Sung, K., Singh, J., & Bridgens, B. (2021). *State-of-the-Art Upcycling Research and Practice: Proceedings of the International Upcycling Symposium 2020*. Springer.

Sung, K., Singh, J., Bridgens, B., & Cooper, T. (2021). Introduction: State-of-the-art upcycling research and practice. In K. Sung, J. Sing, & B. Bridgens (Eds.), *State-of-the-Art Upcycling Research and Practice: Proceedings of the International Upcycling Symposium 2020* (pp. 1–6). Springer.

Sung, K., & Yoon, J. (2019). *Embedding sustainability in design education: The case of design project on systemic changes for sustainable businesses based on upcycling* [conference paper]. Expanding Communities of Sustainable Practice Symposium 2018, Leeds, UK.

Sung, K., & Yoon, J. (2021). Designing for positive upcycling experiences with people's well-being in mind. In K. Sung, J. Sing, & B. Bridgens (Eds.), *State-of-the-Art Upcycling Research and Practice: Proceedings of the International Upcycling Symposium 2020* (pp. 135–139). Springer.

PART III
ENERGY SECTOR

8

FROM GEOLOGIST TO SOLAR ENERGY SCIENTIST

Pathways to improving the performance and sustainability of photovoltaics

Patrick Isherwood

Early adventures in geoscience

I have always had a particular fascination for rocks, and especially gemstones. As a child, family holidays were usually to Scotland, and collecting gems wherever we went was both enjoyable and exciting: tiny garnets from Kinloch Hourn, massive citrine from Loch Lomond, crystal-clear smoky quartz and weathered cubes of pyrite from Ben Lawyers, serpentinised marble from Iona, agates from Mull and Culzean. It was apparent from a very young age that I would become a geologist, and after my A-levels, I undertook an MSci in Geoscience at the University of Durham.

From rocks to sunlight

How does a geologist become a solar energy researcher? As with most degrees, geology can lead to many possible career paths, but for geology, three industries in particular are obvious choices. The oil industry employs many geoscientists in numerous roles. As a recent graduate however, this did not appeal, both for ethical reasons and because I found the geology associated with oil to be dull. Despite also being ethically dubious, I considered mining more seriously, but I was keen to remain in the UK, where it is no longer a major industry. This left geotechnical engineering. I graduated in 2008, just as the entire global financial edifice collapsed into a heap of smouldering ashes and shrill recriminations. The crash had knock-on effects in many industries, and the geotechnical sector suffered along with the rest. Job-hunting is a soul-destroying business even when times are good, but collecting increasing numbers of "no thank you" letters – or more precisely, "thanks, but we have just cancelled our graduate programme" letters – is

DOI: 10.4324/9781003380566-12

particularly depressing. I therefore applied for and was accepted onto an MSc programme at Newcastle to study Engineering Geology. I felt that this should at least give me some relevant training and education in lieu of hands-on experience in the industry. In this, I believe I was correct since most of my course-mates went on either to work in the industry or to study for PhDs in the wider geotechnical field. While I enjoyed the course, however, I rapidly came to realise that I was not fond of the subject. This was a significant problem since I had apparently run out of plausible alternatives.

Whilst the winter during my MSc was particularly hard, with unusually low temperatures and snow lasting well into the new year, Easter was beautiful, with plentiful warmth and sunshine. The world took on the improbable and almost unnaturally vibrant green appearance of full spring. I contemplated a comment an otherwise long-forgotten teacher had made when I was still at school: "If we could fully understand and replicate how photosynthesis works, we could concrete over the world". Consideration of ethics, practicalities, and the general undesirability of such an approach aside, it is interesting that plants have evolved an effective means for converting sunlight into a more useful and usable form of energy. If we could harness that or possibly mimic some part of the process, we could more readily, and in particular more cheaply, harvest what at that point seemed to be an almost unlimited energy supply. My conversion from geologist to solar energy researcher had begun.

I completed my MSc – I am nothing if not stubborn – but during the remainder of that year and over the following year, I began to implement a research programme to see if I could come up with a means of converting light to electricity using the photosynthetic light reactions, or some equivalent process. As an individual, my laboratory tools and reagents were minimal. Until my MSc graduation, however, I still had access to university journal subscriptions, which significantly helped develop my understanding of photovoltaics. It rapidly became clear that the particular type of device I had initially envisaged is known as a dye-sensitised or dye cell, sometimes also referred to as a Grätzel cell after one of the original inventors (Akila, Muthukumarasamy, & Velauthapillai, 2019; Grätzel, 2001). Dye cells involve a photoactive dye, often an organic material with a metal ion bound in the centre, which is absorbed into a mesoporous intrinsic semiconductor material, usually titanium dioxide. The dye intercepts and absorbs visible light, forming excitons or combined electron-hole pairs. These are separated through the injection of the electron into the semiconductor material, from where it is transferred to a transparent conductive top contact and extracted from the cell as useful current (Grätzel, 2001). A back contact provides a return current, which interacts with an electrolyte. This was typically an iodine/iodine salt solution, with iodine ions being oxidised to triiodide by the dye material and triiodide being reduced back to iodine ions at the counter electrode (Joly et al., 2014). More recent design iterations have replaced the iodine-triiodide liquid electrolyte with solid-state organic hole conductors, particularly a material known as Spiro-OMeTAD (Kim et al., 2012).

My initial experiments revolved around the extraction of chlorophyll from plant leaves using surgical spirit and attempts to use it as a photoactive dye. Chlorophyll was chosen because it is the primary photo-absorber material used in plants, so there is ample evidence that it works. I did not initially think I had access to anything I could use as a substitute for the mesoporous metal oxide semiconductor material, however. I then discovered that there was a notable interest in using iron sulphide as a cheap and environmentally friendly absorber material within the academic community (Nakamura & Yamamoto, 2001). Iron sulphide, FeS_2, has a band gap of around 0.95 eV, which is reasonably close to the nominal ideal for a solar absorber (Le Donne, Trifiletti, & Binetti, 2019). It is also a naturally occurring mineral known as pyrite, of which I had a plentiful supply thanks to my penchant for collecting rocks as a child. My first working solar cell was constructed using thin copper wires for both front and back contacts, a layer of finely ground pyrite soaked in chlorophyll solution, and a filter paper spacer soaked in salty water to act both as the electrolyte and to keep the front and back contact wires from touching. The structure was held between two microscope slides and cemented together using waterproof glue to prevent drying. Although hardly the most aesthetically pleasing device, it did produce both measurable voltage and current when exposed to sunlight, and a very rough calculation indicated that it had an efficiency of around half a percent.

From this point on, I began to investigate ways of improving some of the specific parts of the cell. Several papers suggested the use of starch-based plastic as a solid-state electrolyte, using metal halides such as sodium chloride to increase the conductivity and glycerine as a plasticiser (e.g., Finkenstadt & Willett, 2004; Ma, Yu, & He, 2006). This proved to be a relatively simple material to produce, with measurable conductivities similar to those reported in the literature. It also had the benefit of being optically transparent, possibly providing a means of enabling current extraction without copper wires. In particular, however, I looked for better absorber materials. A vital issue with chlorophyll is that it absorbs light in the red and blue parts of the spectrum but does not absorb anything through the yellow-green region, where most of the light typically occurs. This severely limits its value as a photo-absorber material if a high-efficiency device is desired. Chlorophyll is also sensitive to light, oxygen, and acids, and degrades over time. This is not a significant problem in plants since they can produce more, but it is a major drawback in a solar cell. Two materials in particular showed promise as alternatives: ferric tannate, or iron-gall ink, and an iron-chlorophyll reaction product I discovered by accident. Ferric tannate is a well-known blue-black insoluble material formed by reacting a tannin solution with an iron salt (Jančovičová, Čeppan, Havlínová, Reháková, & Jakubíková, 2007). As an ink, it is typically prepared using an iron II salt since ferrous tannate is water-soluble. However, it will oxidise to the insoluble ferric (iron III) form as the ink dries (Jančovičová et al., 2007). Although largely superseded by modern inks, it has seen extensive use in the past, with the majority of historical documents written using it, and consequently it has an excellent stability record – arguably better than that of silicon (Jančovičová et al., 2007). Initial experiments demonstrated that ferric tannate would work as an absorber, but producing it in film form proved to be extremely challenging. I decided to try mixing ferric tannate with chlorophyll to improve the performance of cells made using both absorbers, and discovered that mixing my iron III salt (homemade ferric acetate) with chlorophyll solution produced a spectacular deep black ink. This proved to be more effective as an absorber than either chlorophyll or ferric tannate, and I managed to make a device with an approximate efficiency of one percent. Unfortunately, despite several attempts under more controlled conditions, I have not been able to repeat this experiment, so what exactly this material was remains to be discovered. Having successfully made several working devices, I decided to investigate the possibility of undertaking a PhD, enabling me to continue my research in a proper laboratory and with better access to materials and equipment. With this in mind, I approached the photovoltaic materials group at the Centre for Renewable Energy Systems Technology, Loughborough University, where I was offered a position as a fully funded doctoral candidate.

PhD research: transparent conducting oxides

From the beginning of my work with photovoltaics, I was particularly fascinated by a unique problem: how to extract current without shading the device. In silicon cells, this is mainly achieved by compromise – the silver filaments, fingers, and busbars, which are typically seen on the top of the cell, allow for current extraction but shade small areas of the device, reducing the active area (Faes et al., 2014; Luque & Hegedus, 2011). In most thin-film cells, a transparent conducting oxide (TCO) is used as the top contact (Calnan & Tiwari, 2010; Luque & Hegedus, 2011). My PhD provided the opportunity to study these materials further. The top contact in a thin film solar cell should ideally be transparent and electrically conductive, allowing light into the device whilst simultaneously enabling current extraction. There are several n-type materials which are suitable for this use, and which have been known for several decades, such as tin-doped indium oxide (ITO), fluorine-doped tin oxide (FTO), and aluminium-doped zinc oxide (AZO) (Calnan & Tiwari, 2010; Gerhardinger & Strickler, 2008; Luque & Hegedus, 2011). One part of my thesis involved a study of AZO, but for the most part, I was interested in p-type materials. These are interesting because so far, no one has identified a commercially viable p-type metal oxide that is transparent to visible light and sufficiently conductive to be useful. The first p-type oxide to be identified, which had some degree of transparency, and which demonstrated a conductivity slightly higher than that of a typical insulator, was copper aluminium oxide. This is a delafossite-structured material (Kawazoe et al., 1997; Yanagi et al., 2000). Subsequent investigations demonstrated that other delafossites also displayed some degree of transparency and p-type conductivity, and it soon became apparent that copper was the key (Yanagi et al., 2000). Most of my PhD followed on from this work, looking at copper oxide-based materials. In particular, I investigated cupric oxide, copper tin oxide, and copper zinc oxide.

Under most circumstances, metal oxides can readily be made n-type, and there are any number of n-type transparent conductors, including those already mentioned. In addition to photovoltaics, these materials have many uses, such as flatscreen TVs and computer monitors, touchscreens, smart windows, and double glazing. The reason why n-type metal oxide semiconductors are so much more prevalent is a function of the typical metal oxide band structure. In most metal oxides, the conduction band is formed from the outer electron shells of the metal cations. Because of the atomic sizes and moderate electronegativity of many metals, the resulting conduction band can be both topographically varied and energetically broad, so it is relatively easy for electrons to move within it (Hautier, Miglio, Ceder, Rignanese, & Gonze, 2013; Yanagi et al., 2000). The valence band, however, is usually made up of oxygen $2p$ states. Oxygen is very electronegative, and the $2p$ state is close to the atomic nucleus, resulting in a valence band which is both energetically narrow and topographically flat (Hautier et al., 2013; Isherwood, Butler, Walsh, & Walls, 2015; Kudo, Yanagi, Hosono, & Kawazoe, 1998; Yanagi et al., 2000). Consequently, electron holes experience high localisation forces, resulting in high hole effective masses and low p-type conductivity. Metal cations with full or essentially complete outer d subshells, such as copper, can improve this a little since the d subshell can overlap with the oxygen $2p$ states, causing increased spreading and undulation in the valence band (Hautier et al., 2013; Isherwood et al., 2015; Kudo et al., 1998; Yanagi et al., 2000). This is why both standard copper oxides are p-type and why copper-based oxide materials can exhibit p-type behaviour. Even so, the achievable hole mobilities are low. This can be partially counteracted through very high levels of dopants and advanced processes such as modulation doping (Isherwood & Walls, 2014). During my doctoral research I examined both of these, doping cupric oxide with sodium and building multi-layered thin film stacks of sodium-doped cupric oxide and tin monoxide. Both approaches produced limited success, giving films with conductivities in order of 50 Siemens/cm. However, this is well short of the ~7,000 Siemens/cm achievable using n-type materials. Ultimately, I too failed, and the outcome of my thesis is most readily summarised as "do not try this, it does not work".

Evolving interests and current research

After my doctorate I became a post-doctoral research associate, working on transparent conducting oxides as electrical contacts for organometal halide perovskite devices, before gaining a lectureship at Loughborough. My current research includes thermodynamic modelling of photovoltaics under different spectral conditions, developing low-cost absorber materials, and recycling solar modules.

An important aspect of solar cell design is that installed devices must operate under highly variable and constantly fluctuating conditions. In laboratory testing, solar cells are measured using a solar spectrum known as AM1.5G, or air mass one point five global (BSI Standards Publication, 2019). The term "air mass" denotes the relative thickness of the atmosphere that sunlight travels through to reach the Earth's surface, with an air mass of one meaning that the sun is directly overhead. An air mass of one point five equals to a sun position of 48.2 degrees from the vertical (Isherwood, Cole, Smith, & Betts, 2021). A global spectrum is one that includes diffuse and reflected light, and not just the direct beam component. The officially defined AM1.5G spectrum gives an incident energy value of 1000 W/m^2 (BSI Standards Publication, 2019). Nowhere experiences this level of irradiance all the time, and many places rarely experience it at all (Isherwood et al., 2021). An AM1.5G spectrum is typically seen less than one percent of the year in Loughborough, with typical irradiance values being notably lower even in summer. This is partly because the UK is at a relatively high latitude and typically experiences considerable cloud cover. This would not be a concern if changes in irradiance values could be applied equally across the entire spectrum. The energy balance in any given spectrum, however, depends not only on the time of day and year but also on parameters such as the thickness and type of any clouds, degree of haze, pollution, aerosols and dust, the type and reflectiveness of the surroundings (albedo), and the presence and position of any sources of shade such as buildings, trees or topographical features. In order to understand how a device will perform in any given location, the local variation in spectrum across the typical year must be understood (Isherwood et al., 2021).

This is further complicated by the complex behaviour of solar cells, driven by the changes in intrinsic loss mechanisms resulting from spectral variation. Intrinsic losses are defined as those which are thermodynamically unavoidable. For a standard single junction solar cell under an AM1.5G spectrum, these losses are: below gap transmission, photon cooling (thermalisation), Boltzmann or etendue expansion loss, Carnot loss, and emission loss (Figure 8.1) (Dupré, Vaillon, & Green, 2016; Hirst, 2012; Hirst & Ekins-Daukes, 2011; Isherwood et al., 2021). The first two are a consequence of solar cells being semiconductor devices. The absorber material's band gap determines which parts of the spectrum the cell can absorb, and any light with photon energy below this limit cannot contribute to the cell current. This is the below gap

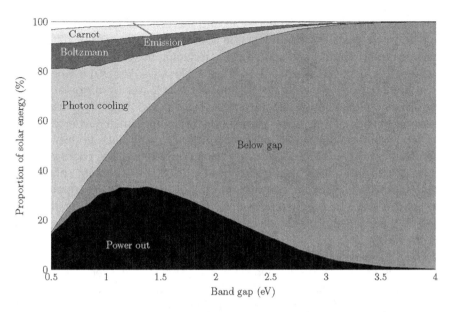

FIGURE 8.1 Intrinsic losses and maximum power output for band gaps from 0.5 to 4 eV, under an AM1.5G spectrum

loss. In addition, the cell can only use the energy in any given absorbed photon equivalent to the band gap, with any additional energy being converted into heat. This is photon cooling or thermalisation. Of the remaining losses, the Carnot loss is a direct result of a solar cell being an unusual form of heat engine, so the absolute thermodynamic conversion limit is a function of the temperature difference between the cell and the sun (Markvart, 2008). Because this temperature difference is enormous, the loss is comparatively small. In any cell, at the maximum power point (the voltage–current combination which gives maximum power output), there is a small probability that any given electron which has been excited across the band gap through light absorption will spontaneously release this energy and combine with an electron hole in the absorber material's valence band. This process is known as recombination. In practice, the energy released by this process is partly released as heat (a form of non-radiative recombination) and partly as light (radiative recombination). In a theoretically perfect cell, however, only radiative recombination is unavoidable (Hirst & Ekins-Daukes, 2011; Isherwood et al., 2021). Finally, the Boltzmann loss results from the optical expansion of the solid angle of light emitted by the sun and re-emitted by the cell through radiative recombination. Because the sun is a very long way from the Earth, the solar solid angle is very small, whilst that of the cell is large. If the cell is horizontal and optically unconstrained, it can emit light into the entire celestial hemisphere. This angle can be reduced by placing mirrors or other reflective surfaces around the cell, reflecting emitted light into the cell. Because the Boltzmann loss is a function of the difference between the solar and cell solid angles, reducing the cell's emission angle reduces this loss (Dupré et al., 2016; Hirst & Ekins-Daukes, 2011; Isherwood et al., 2021). Solar concentrator systems such as lenses or mirrors increase the amount of light hitting the cell and reduce this loss. This is because these systems, in effect, increase the solar solid angle. Unfortunately, the cell must be mounted on a solar tracking system in both cases since both methods eliminate a portion of the diffuse spectrum equivalent to the change in the solar or cell solid angle (Isherwood et al., 2021). At maximum solar concentration or cell angular constriction (46,200 times), the Boltzmann loss disappears entirely, but the cell is only capable of using direct beam light, and no diffuse light is useable at all (Hirst & Ekins-Daukes, 2011; Isherwood et al., 2021).

The remaining spectral energy gives the cell maximum performance value, commonly known as the detailed balance or Shockly-Quiesser limit (Hirst & Ekins-Daukes, 2011; Shockley & Queisser, 1961). The losses and performance can be calculated for different spectra and device band gaps. Understanding and being able to model these loss mechanisms under varied spectra provides a more nuanced picture of the possible energy yield achievable for any given location. This can be improved further by using device models which describe the specific optical behaviour of cell technologies, allowing for better comparison and tailoring of device construction to maximise performance (Isherwood et al., 2021).

More recently, I have developed a keen interest in photovoltaic module recycling. This is an increasingly important subject because of the near-exponential increase in module deployment over the last decade or more (Heath et al., 2020). At present, solar modules are predominantly made using silicon wafer-based cells connected together. Modules have a laminated structure, usually consisting of a glass cover plate, ethylene vinyl acetate (EVA) adhesive, silicon cells and interconnects, another EVA layer, and a fluoropolymer-based composite backsheet (Czanderna & Pern, 1996; Heath et al., 2020). The structure is typically held in an aluminium frame for additional protection and strength. Other designs include the replacement of the polymer backsheet with a second glass plate to form a module which can receive light from both sides, known as a bifacial module, or replacement of the glass with a polymer cover layer. All current modules contain polymeric materials, with the majority using EVA as the adhesive and weather seal. This is because it has been extensively tested and demonstrated to work effectively over the expected module lifetimes of twenty-five years or more and under the harsh outdoor environments that modules experience. This poses a significant problem from a recycling perspective since the polymeric materials used in solar modules, and in particular EVA, are not readily recyclable. Until recently, module design has focused on ensuring long-term stability and weather-proofing, with little regard for end-of-life processing and recycling.

Although still a relatively young area of research, several methods exist for recycling the non-polymeric materials in solar modules. This includes both silicon modules and the two standard thin-film technologies, cadmium telluride and copper indium gallium diselenide (Isherwood, 2022). Existing recycling processes for all types of solar modules involve either burning or pyrolysing the polymeric components, or using some means of separating them from the other module parts and sending them to a landfill (Isherwood, 2022). Neither of these options is satisfactory, and both have notable environmental concerns. Burning and pyrolysis release carbon dioxide and can produce toxic waste gases, particularly if the fluoropolymer backsheet is burned, whilst plastics in landfill constitute a long-term problem due to their slow breakdown rate and a high chance of accidental environmental release. Therefore, it is essential to identify means for removing all plastics from use in modules. This is not simple since polymers – particularly EVA – are used as the adhesive holding everything together. Although dramatic reduction in polymer use has been achieved through the production of edge-sealed glass-backed panels, complete removal of polymers has not yet been achieved (Reinwand et al., 2019). A key area of my ongoing research is into methods of physical or chemical welding of glass sheets to enable the construction of modules requiring no polymeric binders.

Embedded sustainability: future planning

An enduring problem and cause for concern is that much of the ongoing research in the photovoltaics field is limited in scope and ambition. Photovoltaics are an increasingly cheap means of electricity generation, but being cheaper to install than fossil fuel technologies is insufficient. Climate change is a long-term existential threat to our species and to life on this planet, and will require extraordinary measures to combat. We are already doing too little too late, perennially delaying necessary changes because it is politically or economically expedient to ignore the problem. By the time it becomes obvious to even the most intransigent individuals that these changes are needed, it will be far too late. However, there may be ways of dealing with this issue. The development of renewable energy technologies designed from the ground up to be comprehensively recycled in a fully circular fashion, and which are dramatically cheaper than the alternatives, would provide an undeniable economic argument for their rapid deployment, whilst reducing reliance on virgin resources. Technologies that enable a significant reduction in electricity costs would most likely be welcomed by everyone, regardless of their opinions and beliefs regarding anthropogenic climate change. Since renewable resources such as solar, wind and water power are typically relatively dispersed, there is also a strong argument for their use to reduce reliance on energy from geopolitically unstable or unpredictable locations, thereby improving energy security. In order to achieve these goals, future research in this area needs to be much more ambitious. Ultimately, we need to make renewable energy technologies so cheap that they become the only economically plausible option.

References

Akila, Y., Muthukumarasamy, N., & Velauthapillai, D. (2019). TiO2-based dye-sensitized solar cells. In *Nanomaterials for Solar Cell Applications* (pp. 127–144). Elsevier Inc. https://doi.org/10.1016/B978-0-12-813337-8.00005-9.

BSI Standards Publication. (2019). Photovoltaic devices part 3: Measurement principles for terrestrial photovoltaic (PV) solar devices with reference spectral irradiance data (IEC 60904–60903:2019).

Calnan, S., & Tiwari, A. N. (2010). High mobility transparent conducting oxides for thin film solar cells. *Thin Solid Films*, 518(7), 1839–1849. https://doi.org/10.1016/j.tsf.2009.09.044.

Czanderna, A. W., & Pern, F. J. (1996). Encapsulation of PV modules using ethylene vinyl acetate copolymer as a pottant: A critical review. *Solar Energy Materials and Solar Cells*, 43(2), 101–181. https://doi.org/10.1016/0927-0248(95)00150–00156.

Dupré, O., Vaillon, R., & Green, M. A. (2016). A full thermal model for photovoltaic devices. *Solar Energy*, 140, 73–82. https://doi.org/10.1016/j.solener.2016.10.033.

Faes, A., Despeisse, M., Levrat, J., Champliaud, J., Badel, N., Kiaee, M., ... Ballif, C. (2014). Smartwire solar cell interconnection technology. In *Proceedings of the 29th European Photovoltaic Solar Energy Conference and Exhibition* (pp. 2555–2561). https://doi.org/10.4229/EUPVSEC20142014-5DO.16.3.

Finkenstadt, V. L., & Willett, J. L. (2004). Electroactive materials composed of starch. *Journal of Polymers and the Environment*, 12(2), 43–46. https://doi.org/10.1023/B.

Gerhardinger, P., & Strickler, D. (2008). Fluorine doped tin oxide coatings – over 50 years and going strong. *Key Engineering Materials*, 380, 169–178.

Grätzel, M. (2001). Photoelectrochemical cells. *Nature*, 414, 338–344. https://doi.org/10.1038/35104607.

Hautier, G., Miglio, A., Ceder, G., Rignanese, G.-M., & Gonze, X. (2013). Identification and design principles of low hole effective mass p-type transparent conducting oxides. *Nature Communications*, 4. https://doi.org/10.1038/ncomms3292.

Heath, G. A., Silverman, T. J., Kempe, M., Deceglie, M., Ravikumar, D., Remo, T., ... Wade, A. (2020). Research and development priorities for silicon photovoltaic module recycling to support a circular economy. *Nature Energy*, 5(7), 502–510. https://doi.org/10.1038/s41560-020-0645-2.

Hirst, L. C. (2012). Principles of solar energy conversion. In A. Sayigh (Ed.), *Comprehensive Renewable Energy* (Vol. 1, pp. 293–313). Elsevier Ltd. https://doi.org/10.1016/B978-0-08-087872-0.00115-3.

Hirst, L. C., & Ekins-Daukes, N. J. (2011). Fundamental losses in solar cells. *Progress in Photovoltaics: Research and Applications*, 19, 2752–2756. https://doi.org/10.1002/pip.1024.

Isherwood, P. J. M. (2022). Reshaping the module: The path to comprehensive photovoltaic panel recycling. *Sustainability*, 14(3), 1676. https://doi.org/10.3390/su14031676.

Isherwood, P. J. M., Butler, K. T., Walsh, A., & Walls, J. M. (2015). A tunable amorphous p-type ternary oxide system: The highly mismatched alloy of copper tin oxide. *Journal of Applied Physics*, 118(10). https://doi.org/10.1063/1.4929752.

Isherwood, P. J. M., Cole, I. R., Smith, A., & Betts, T. R. (2021). The impact of spectral variation on the thermodynamic limits to photovoltaic energy conversion. *Solar Energy*, 221, 131–139. https://doi.org/10.1016/j.solener.2021.04.037.

Isherwood, P. J. M., & Walls, J. M. (2014). Cupric oxide-based p-type transparent conductors. *Energy Procedia*, 60(May), 129–134. https://doi.org/10.1016/j.egypro.2014.12.354.

Jančovičová, V., Čeppan, M., Havlínová, B., Reháková, M., & Jakubíková, Z. (2007). Interactions in iron gall inks. *Chemical Papers*, 61(5), 391–397. https://doi.org/10.2478/s11696-007-0053-0.

Joly, D., Pellejà, L., Narbey, S., Oswald, F., Chiron, J., Clifford, J. N., … Demadrille, R. (2014). A robust organic dye for dye sensitized solar cells based on iodine/iodide electrolytes combining high efficiency and outstanding stability. *Scientific Reports*, 4, 1–7. https://doi.org/10.1038/srep04033.

Kawazoe, H., Yasukawa, M., Hyodo, H., Kurita, M., Yanagi, H., & Hosono, H. (1997). P-type electrical conduction in transparent thin films of CuAlO2. *Nature*, 389, 939–942.

Kim, H. S., Lee, C. R., Im, J. H., Lee, K. B., Moehl, T., Marchioro, A., … Park, N. G. (2012). Lead iodide perovskite sensitized all-solid-state submicron thin film mesoscopic solar cell with efficiency exceeding 9%. *Scientific Reports*, 2, 1–7. https://doi.org/10.1038/srep00591.

Kudo, A., Yanagi, H., Hosono, H., & Kawazoe, H. (1998). SrCu2O2: A p-type conductive oxide with wide band gap. *Applied Physics Letters*, 73(2), 220. https://doi.org/10.1063/1.121761.

Le Donne, A., Trifiletti, V., & Binetti, S. (2019). New earth-abundant thin film solar cells based on chalcogenides. *Frontiers in Chemistry*, 7. https://doi.org/10.3389/fchem.2019.00297.

Luque, A., & Hegedus, S. (Eds.). (2011). *Handbook of Photovoltaic Science and Engineering* (2nd ed.). John Wiley & Sons.

Ma, X., Yu, J., & He, K. (2006). Thermoplastic starch plasticized by glycerol as solid polymer electrolytes. *Macromolecular Materials and Engineering*, 291(11), 1407–1413. https://doi.org/10.1002/mame.200600261.

Markvart, T. (2008). Solar cell as a heat engine: Energy-entropy analysis of photovoltaic conversion. *Physica Status Solidi (A) Applications and Materials Science*, 205(12), 2752–2756. https://doi.org/10.1002/pssa.200880460.

Nakamura, S., & Yamamoto, A. (2001). Electrodeposition of pyrite(FeS2) thin films for photovoltaic cells. *Solar Energy Materials and Solar Cells*, 65(1), 79–85. https://doi.org/10.1016/S0927-0248(00)00080-00085.

Reinwand, D., King, B., Schube, J., Madon, F., Einhaus, R., & Kray, D. (2019). Lab-scale manufacturing of medium-sized N.I.C.E.™ modules with high-efficiency bifacial silicon heterojunction solar cells. *AIP Conference Proceedings*, 2156(September). https://doi.org/10.1063/1.5125874.

Shockley, W., & Queisser, H. J. (1961). Detailed balance limit of efficiency of p-n junction solar cells. *Journal of Applied Physics*, 32(1961), 510–519. https://doi.org/10.1063/1.1736034.

Yanagi, H., Inoue, S., Ueda, K., Kawazoe, H., Hosono, H., & Hamada, N. (2000). Electronic structure and optoelectronic properties of transparent p-type conducting CuAlO2. *Journal of Applied Physics*, 88(7), 4159–4163. https://doi.org/10.1063/1.1308103.

9

FROM FACING ENDLESS UNSCHEDULED POWER OUTAGES TO CLEAN ENERGY TECHNOLOGIES

Research journey to zero-carbon future

Prabhakaran Selvaraj

Early days: poultry farm dream

I grew up at a farmhouse in southern India, where weekends and school holidays were mainly spent helping my parents with farming and playing gully cricket at other times. In that part of the world, academic degrees and higher education were considered tools to get white-collar jobs and a comfortable life. I desired to own a poultry farm as it can be a fulfilling and profitable venture. However, I realised I was too young to start one. So I decided to study for a degree. Physics was my favourite subject at school. After finishing a higher secondary course (Year 12), I moved to the city of Coimbatore and undertook a BSc in Physics at Kongunadu Arts and Science College, Bharathiar University.

Plenty of sunlight but no power

Sugarcane is a major cash crop in South India, particularly in states like Tamil Nadu, Karnataka, and Maharashtra. The cultivation of sugarcane requires a lot of water, and farmers usually rely on groundwater or surface water sources for irrigation. During summer, when the demand for irrigation water is high, farmers tend to operate their pumps for longer hours, which puts a strain on the power supply. This, coupled with the already limited power capacity in the region, leads to frequent power outages. From 2007 to 2010, when I was finishing school and in the first year of my undergraduate programme, South India faced a series of power outages for various reasons. Some significant power cuts occurred in the summer when electricity demand peaked. These blackouts varied from a few hours to several days, disrupting daily life, business operations, and economic activities. Our family

DOI: 10.4324/9781003380566-13

was affected by this crisis as we had to be awake day and night to irrigate the sugarcane farm whenever power was available to operate the pumps. To address the power crisis, the government of India launched several initiatives, such as promoting renewable energy, building new power plants, upgrading transmission and distribution networks, reducing power theft, and encouraging energy conservation. This encouraged me to learn more about electricity generation and the available renewable energy resources.

Whilst pursuing my master's degree in physics, I had an opportunity to interact with PhD researchers almost daily. Even though all of them were working on PV (photovoltaic) materials development, it helped me to understand available research internships and further opportunities. In the summer of 2013, I was selected for a summer internship at Indira Gandhi Centre for Atomic Research (IGCAR). I was one among 20 students who were selected from all across India for this internship. I attended various research lectures and software training given by the researchers. I gained a lot of experience in material characterisation techniques such as XRD, SEM, TEM, AFM, XPS, Raman spectroscopy, and UV-Vis spectrophotometry. I had an opportunity to work with senior researchers at the materials science division and carried out a project entitled "Compression behaviour of porous gold for electrochemical applications".

The notable feature of porous metallic solids is the continuous three-dimensional pore (free volume) region with interconnected ligament (plate-like dense solid) networks throughout the volume. These porous metallic solids have potential demand in nuclear, chemical and aircraft industries in developing membranes, catalytic supports, electrodes in electrochemical and fuel cells, sensors, energy absorbers, and so on. Their low density, high surface area and weak coordination of atoms at the ligament pore interfaces are the critical factors for achieving mechanically stable porous network solids with externally driven functional properties. A range of alloy specimens such as Cu-Au, Ag-Au, Pb-Au, and Al-Au have been used by the researchers as starting materials to produce porous metallic solids by selective leaching of the more electrochemically active elements from the alloys at an applied electrochemical cell potential of less than 1V. Among the starting materials used, Ag-Au was our primary interest for producing porous Au. The scientific issue that remains unsolved in porous Au is its mechanical behaviour under various external-driven loads. The compression behaviour of porous gold was carried out in the summer project. An Ag-Au alloy was prepared using an arc melter with a suitable chemical composition. By an electrochemical dealloying process, porous Au was fabricated, and its mechanical compression was demonstrated. Finally, I had the opportunity to present all the results at the end of the summer research programme. The experience and research conversations I had with the scientists there were the reasons I started thinking about doing PhD research abroad, as they encouraged me to look for opportunities to pursue my research dream.

For my master's dissertation, I deposited ~250nm thickness cadmium oxide (CdO) thin films on p-type silicon (100) substrates by the chemical bath deposition (CBD) method. n-CdO/p-si (100) heterojunction diodes were fabricated. The structural, morphological, compositional, and defect properties of prepared CdO thin films were characterised using XRD, SEM, EDAX, and PL analysis. Diode characteristics were analysed. This gave me a better understanding of thin film technology and a path to seeking PhD positions in the same discipline. I applied to a few universities and was offered a fully funded doctoral candidate position at the Department of Renewable Energy, University of Exeter.

Transparent dye-sensitised solar cells for building integration: doctoral research

Several factors, including solar cell material (e.g., silicon, semiconductor compounds, sensitiser, electrolyte); cell size (the larger the cell size, the more the individual cells transform into either more voltage or current); and intensity and quality of the light source, determine the amount of electricity generated (Goetzberger et al., 2002). Wafer-based monocrystalline or polycrystalline silicon (mc-Si and pc-Si) and GaAs thin film solar cells belong to the first-generation solar cells. The former two account for 93% of the global market, with their best module efficiencies being around 22% for mc-Si and 16% for pc-Si (Green et al., 2022). Second-generation solar cells are based on thin film technologies such as amorphous silicon (a-Si), cadmium telluride (CdTe), and copper-indium- gallium selenide (CIGS), with thicknesses that are usually 1–2 orders of magnitude lower than those wafer-based crystalline silicon solar cells. As a result of the small amount of active material needed, lower manufacturing costs and shorter energy payback times are achieved. In addition, thin film solar cells can be used with flexible substrates. Best laboratory efficiencies are close to 20% for CdTe and CIGS and 10% for a-Si (Luceño-Sánchez et al., 2019). The category of third-generation solar cells originally comprised technologies that allow for achieving ultra-high efficiencies above the Shockley-Queisser limit. Examples are tandem or multi-junction cells, intermediate band solar cells, hot carrier cells, photon up- or down-conversion, and concentrator solar cells (Yan & Saunders, 2014). Emerging technologies with the potential to significantly reduce the cost per watt peak, such as organic solar cells, dye-sensitised solar cells, and perovskite solar cells, are also typically included in the category of third-generation solar cells (Sulaeman & Zuhairi Abdullah, 2017). These cells are considered a promising technology, especially in building-integrated PV markets for the future, due to their transparency, ease of processing, low production cost, flexibility, and high efficiency, but their long-term stability is not yet high enough to be competitive. However, they have the potential to become more efficient than silicon cells through better light trapping and material selection (Płaczek-Popko, 2017).

The general principle of Building-integrated photovoltaics (BIPV) is that PV modules are integrated into the building envelope, substituting standard glass and other cladding materials with glass/glass laminates encapsulating PV cells within. These BIPV modules generate electricity at the 'point of use', maximising energy efficiency and eliminating transmission losses. The electricity generated using this system is directly fed into the building, making the source of energy the sole point of its consumption. However, they have very tough competition from the standard silicon panels. Any increase in the electricity prices improves the viability of semi-transparent photovoltaic systems. By becoming an integral part of the building architecture, BIPV can enhance building energy efficiency through electricity generation, provide daylight transmission, and improve the thermal properties of building facades (Baig, 2015). BIPV systems can be integrated into different building architectures, including windows, vertical glazing, canopies, greenhouses, curtain walls, roofs, window awnings, etc.

Building-integrated photovoltaic windows (BIPW) are expected to be an innovative glazing technology, which, apart from electricity production, are linked with the building heating or cooling loads and artificial lighting (Jelle, 2016). For glazing applications, semitransparency is a precondition (Saifullah et al., 2016), as natural daylight penetrating through this semi-transparent PV makes the indoor environment comfortable. Available PV types for glazing applications include crystalline silicon, CdTe, a-Si, CIGS, DSSC (dye-sensitised solar cell), and perovskite. Since c-Si cells are typically opaque, there are also important compromises in terms of lighting (shadows in the building interior) and limited external view (Jelle et al., 2012; Shukla et al., 2016; Skandalos & Karamanis, 2015). Regular distribution of opaque c-Si can offer daylighting; however, this structure blocks natural viewing (Ghosh et al., 2019). Thin film second-generation CdTe, a-Si, and CIGS are other options for PV glazing applications (Sun et al., 2018; Wang et al., 2017). With thin film incorporation in a glass–glass construction, commercial products with up to 50% transparency are available in the market. The introduction of this technology provided more homogeneous daylighting of the interior spaces compared to crystalline solar cells. However, light-induced defects, shortage and toxicity of materials used in a-Si, CIGS, and CdTe technologies have limited the opportunity to apply them in glazing applications (Lee & Ebong, 2017).

Moreover, the power conversion efficiency is connected to its visual transmittance, and therefore, extensive performance optimisation should be considered (Cuce, 2016; Saifullah et al., 2016; Skandalos & Karamanis, 2015). Due to their advantages, third-generation dye solar cells are a potential candidate for BIPV applications. While many research groups investigate the working principles of DSSCs and new developments have been achieved concerning their efficiency and large-scale applications, new companies founded in the meantime try to carry DSSC technology into the marketplace, evaluating all process steps needed for industrial production. Considering the knowledge gaps identified in the literature, a few major research questions can be raised: Can DSSCs be used as BIPW material? Can their colour properties be compared to the available commercial glazing materials? How much can optical concentrators improve DSSC performance? How do the charge transfer properties change under high light intensity?

My PhD thesis covers three aspects of DSSC research. Firstly, the operating process of the DSSC is greatly dependent on the individual component properties and the interplay between them, requiring systematic analysis to gain insight into the system's inherent complexity. To achieve a more fundamental understanding of the device mechanisms of dye-sensitised solar cells, three simple and low-cost new materials for cost-effective DSSCs were introduced. The photovoltaic performance of the synthesised mesoporous TiO_2 devices is significantly higher than the commercial P25 titania electrode devices. The highest power conversion efficiency of 6.08% with a photo current density of 12.63 mA/cm^2 was achieved because of enhanced light harvesting, due to the high surface area (Selvaraj et al., 2019).

Additionally, the new m-HRD-1 sensitiser-based device achieved 20% more short-circuit current due to the more effective anchoring of the dye molecules. Furthermore, The CZTS counter electrode DSSC showed a promising power conversion efficiency of 3.91%. Even though this is lower than the standard platinum electrode device, the cost effectiveness of this electrode makes it a potential candidate for large-scale DSSC applications (Kirubakaran et al., 2018). As is evident from the results, it is necessary to understand the specific factors and parameters affecting the photovoltaic performance to achieve highly efficient and low-cost DSSCs. In order to investigate DSSC glazing's (Figure 9.1)

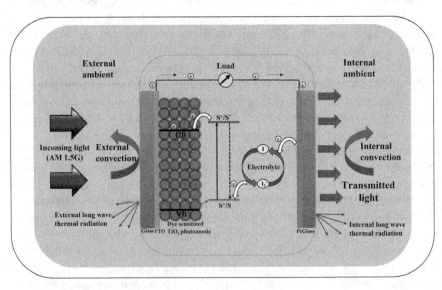

FIGURE 9.1 Schematic representation of DSSC glazing

colour and glazing properties under different environmental factors, semi-transparent DSSCs were fabricated and analysed. It was found that the transparent DSSCs offer only 2.7% lower colour rendering index (CRI) and correlated colour temperature (CCT) values than the vacuum and double-glazing (Ghosh et al., 2018). In southwest UK (Penryn, UK – 50.16° N, 5.10° W), the solar factor was higher in January than in July. Moreover, 21% more glares can be reduced on a clear sunny day than double glazing using 37% transparent and 6% power conversion efficient DSSC glazing (Selvaraj, Ghosh et al., 2019). The results showed that semi-transparent DSSCs are a potential glazing system for either new or retrofitted windows. Building engineers and architects will be able to benefit from these results to design a new low-energy or retrofit building with DSSC glazing.

A key parameter limiting the amount of electricity generation is the area of solar cells used. By concentrating the incoming sunlight, a larger amount of electricity can be generated using fewer solar cell materials. Incorporating the concentrating photovoltaics into any part of the building architecture is called building-integrated concentrating photovoltaics (BICPV) (Baig et al., 2015). A low solar concentrator was introduced for DSSC to increase incident light's intensity. The photovoltaic performance of the bare semi-transparent devices was studied, and the relationship between the transparency and PV performance was understood. The devices' photovoltaic performance increases with the photoanode's thickness until a certain point. Then, for high-thickness devices, it starts decreasing, due to the long electron diffusion length. The overall performance of the concentrator-coupled devices was more than 50% higher than the bare DSSCs. Interestingly, an increase of 67% in power conversion efficiency was observed at 36°C for the concentrator-coupled device under one sun illumination. Moreover, under different device operating temperatures, the low concentrator system coupled device's stability was similar to the bare unconcentrated device. From the impedance studies, it was found that high light intensity reduces the interfacial resistances and improves the performance of semi-transparent DSSCs (Selvaraj, Baig, Mallick, & Sundaram, 2018; Selvaraj, Baig, Mallick, Siviter, et al., 2018).

All the findings in my PhD thesis provide valuable insights into solving the scaling-up problem of DSSCs using solar concentrators. Since the effect of device temperature does not significantly impact on the performance, a long-term stable DSSC glazing system could replace double-glazing in the future. However, further work is needed to unlock the full potential of DSSCs.

Experience in energy industry and research

Since I had no work experience, I decided to gain some by working in the energy industry after completing my PhD. I became an energy modelling analyst to develop energy and financial models using electric and gas data gathered from sites in the UK, performing carbon emission calculations using all the energy data from the existing power station fleet to develop generation forecast models to install microturbine-based combined heat and power (CHP) solutions. This, also known as cogeneration, produces electricity and thermal energy on-site, replacing or supplementing electricity provided by a local utility, with fuel burned in an on-site boiler or furnace. CHP systems increase energy security by producing energy at the point of use, and significantly improve energy efficiency (Murugan & Horák, 2016). Even though not much research was involved in that job, I enjoyed the role and learnt a lot more about the energy industry in the UK. After about 18 months, COVID-19 significantly impacted businesses in the UK, and I was among a few employees to be laid off by my employer. I therefore applied for postdoctoral research positions and was offered a post in the same research group in which I did my PhD.

My EPSRC-funded research at the University of Exeter involved designing, developing, and characterising novel optically integrated glass-on-glass photovoltaic systems for buildings to provide better insulation as well as indoor luminous insulation, and to generate electricity. One of the potential solutions for lowering the overall installation cost of a building-integrated photovoltaic (BIPV) system has been identified as the low-concentration photovoltaic (LCPV) system. A dielectric-based linear concentrating element was designed with a geometrical concentration ratio of 2.4×, intended to operate under extreme incident angles of 0° and 70°. This solar concentrator was attached to a 72 cm^2 passivated emitter and rear contact silicon solar cell module. Under standard test conditions, it was found that the linear concentrator increases the module's short circuit current by 2.3× and the maximum power by 2.1× when compared with a bare solar cell module. A scaled-up building-integrated photovoltaic window system (9 modules of 72 cm^2) was developed to monitor its outdoor long-term performance.

I was also involved in a European Regional Development Fund sponsored research project working on the development of a renewable energy (solar/wind with integrated storage options) system for use on a dairy farm in Cornwall, UK, in order to achieve energy independence. This included recording energy demand on a test farm site as well as the available renewable energy resource (wind, solar, and biomass), modelling optimum configurations of supply and demand for different scenarios, and developing a resolved model for optimising the candidate renewable energy systems for energy independent farming. The farm's carbon footprint running cost was calculated before and after the installation of renewable energy resources based on the developed time-resolved model. It was found that the new carbon running costs of the site are lower due to the potential use of renewable energy instead of grid electricity and biogas instead of diesel.

Current research and future prospects

Windows are an important source of heat gain or heat loss in buildings. Windows are an assembly of glazing, sash, and frame. Glazing refers to the transparent part of a window, the sash is a casing in which the glass sheets of a window are set, and the frame is the complete structural enclosure of the glazing. Fenestration refers to the design and position of windows in a building or dwelling. The window's thermal performance depends on the number of glass sheets, the space between glass sheets, the emissivity of the coatings on the glass sheet, the frame in which the glass is installed and the type of spacers that separate the sheets of glass (Memon, 2013).

Historically, there has been a step change in the methods of further insulating a single-pane window. Firstly, a significant change was made by adopting a low emittance coating, which reduces the amount of radiant heat transfer. However, in extreme climates, further insulation was developed through the addition of a layer of insulating gas. Now, the double-glazed window is commonplace, where a standard configuration consists of a gas gap of air at a distance of about 12–18 mm. Typically, however, a more desirable insulation is achieved using Argon as the gas in the gap. Increasing the number of panes, and thus air gaps, combining multiple low emittance coatings, and using low thermal conductivity gases such as Krypton or Xenon has led to even higher insulation levels. At the centre of a single pane of glass, air-to-air conductance is about 6 W m^{-2} K^{-1} (this is known as the U value of the glass). The thermal conductance can be lowered to about 0.4 W m^{-2} K^{-1} when using multiple lower thermal conductivity gas gaps and with multiple surface coatings. Nevertheless, there are several issues that limit the potential use of the existing technologies, particularly in all areas of the building envelope, such as roof skylights/windows:

i The existing technologies become thicker and heavier than desired at lower U-values;
ii The service life performance cannot be well defined over 25 or more years;
iii For existing buildings, the required frame changes increase costs; and
iv The production cost and energy embodiment are unacceptably high for the lowest U-value performance.

A unique alternative to the existing gas-filled insulting glazing is vacuum glazing. Vacuum glazing consists of two panes of glass, very much like gas-filled glazing. However, instead of using the gas as the insulating medium, a vacuum is established in the gap, which removes the impact of gaseous conduction and convection processes. The evacuated gap is about a tenth of a millimetre wide and should be at a pressure of 0.1 Pa or lower. To stop atmospheric forces from bringing the glass panes into contact, an array of high-strength spacers is positioned in the gap to maintain a constant separation distance between the glass panes (Eames, 2008). The edges of the panes are sealed hermetically to completely stop gas ingress and keep a stable vacuum level over 25 years or more of service life.

Vacuum glazing has a long history, most of which is documented in the patent literature. It is well known that the first vacuum glazing design, a proof-of-concept product, was reported at the University of Sydney in the early 1990s. Soon after, the first commercial product was produced by Nippon Sheet Glass (NSG), Japan, in 1996. NSG has manufactured several million vacuum glazing units and has shown excellent reliability in many building types. Several major research studies have been undertaken at other academic institutions, in government laboratories, and by other companies. The most critical design condition to satisfy in any vacuum glazing design is that the overall thermal conductance is as low as possible (Cuce & Cuce, 2016). Current commercially available vacuum glazings using annealed glass can have centre-of-glazing U values as low as 0.6 W m^{-2} K^{-1} in a structure 10 mm thick. There is, however, still an issue surrounding the current manufacturing process. At this time, all the commercial manufacturing lines globally which produce commercial products are based on a vacuum glazing design that uses a solder glass edge seal, and therefore a batch or in-line furnace step that requires high temperatures (about 350–450°C) over relatively long periods (2–5 hours). Manufacturing and energy-embodiment cost analysis performed at the University of Sydney have shown a high undesirable impact of utility (energy) use in manufacturing. This results in a vacuum glazing production cost that is two to five times greater than the manufacturing cost of an argon-filled insulated glazing unit product. The current focus of much of the industry and academic research is to increase the throughput of the manufacturing process while reducing the costs. Multiple efforts exist to realise a manufacturing process that can be performed at room temperature. This would significantly reduce the impact of energy use and allow for a product that may still have a higher manufacturing cost than a traditional glazing unit but will allow for much lower costs in the final window installation for a residential or commercial building (Kocer, 2019).

Previously, researchers from Ulster University developed and patented a low-temperature sealing process for the production of vacuum glazing (Eames et al., 2007). The benefit of the low-temperature (less than 200°C) sealing procedure is that it enables the use of tempered glass and a variety of soft low-e coatings (Memon et al., 2015). A measured heat transmittance of 0.86 W m^{-2} K^{-1} was achieved in the central area of the vacuum glazing for two low-e coatings and support pillars of 0.4 mm in diameter. My current research at Loughborough University is to design, develop, and characterise new, durable, low-cost, low-heat loss vacuum glazing and frames integrated with a switchable layer to provide U-values down to 0.5 W m^{-2} K^{-1} along with daylight control. This research aims to investigate the process chemistry and physics to improve in the edge seal design and develop an optimised low-cost seal and fabrication process for vacuum glazing. The developed switching layers will be added to the formed vacuum glazing with an additional glass pane for daylight control.

Combining semi-transparent solar panels with low-temperature processed vacuum glazing providing a service life performance over 25 or more years will improve overall thermal insulation, reduce solar heat gain, let comfortable daylight into the building, and generate green electricity. That could be the key to energy-efficient buildings that help to achieve the zero-carbon future.

References

Baig, H. (2015). *Enhancing Performance of Building Integrated Concentrating Photovoltaic Systems* (Issue February) [University of Exeter]. http://hdl.handle.net/10871/17301.

Baig, H., Sellami, N., & Mallick, T. K. (2015). Performance modeling and testing of a building integrated concentrating photovoltaic (BICPV) system. *Solar Energy Materials and Solar Cells*, 134, 29–44. https://doi.org/10.1016/j.solmat.2014.11.019.

Cuce, E. (2016). Toward multi-functional PV glazing technologies in low/zero carbon buildings: Heat insulation solar glass – latest developments and future prospects. *Renewable and Sustainable Energy Reviews*, 60, 1286–1301. https://doi.org/10.1016/j.rser.2016.03.009.

Cuce, E., & Cuce, P. M. (2016). Vacuum glazing for highly insulating windows: Recent developments and future prospects. *Renewable and Sustainable Energy Reviews*, 54, 1345–1357. https://doi.org/10.1016/j.rser.2015.10.134.

Eames, P. C. (2008). Vacuum glazing: Current performance and future prospects. *Vacuum*, 82(7), 717–722. https://doi.org/10.1016/j.vacuum.2007.10.017.

Eames, P. C.*et al.* (2007). *Method of Sealing Glass – United States Patent, US 7,204,102 B1.*

Ghosh, A., Selvaraj, P., Sundaram, S., & Mallick, T. K. (2018). The colour rendering index and correlated colour temperature of dye-sensitized solar cell for adaptive glazing application. *Solar Energy*, 163. https://doi.org/10.1016/j.solener.2018.02.021.

Ghosh, A., Sundaram, S., & Mallick, T. K. (2019). Colour properties and glazing factors evaluation of multicrystalline based semi-transparent Photovoltaic-vacuum glazing for BIPV application. *Renewable Energy*, 131, 730–736. https://doi.org/10.1016/j.renene.2018.07.088.

Goetzberger, A., Luther, J., & Willeke, G. (2002). Solar cells: past, present, future. *Solar Energy Materials & Solar Cells*, 74.

Green, M. A., Dunlop, E. D., Hohl-Ebinger, J., Yoshita, M., Kopidakis, N., Bothe, K., Hinken, D., Rauer, M., & Hao, X. (2022). Solar cell efficiency tables (Version 60). *Progress in Photovoltaics: Research and Applications*, 30(7), 687–701. https://doi.org/10.1002/pip.3595.

Jelle, B. P. (2016). Building integrated photovoltaics: A concise description of the current state of the art and possible research pathways. *Energies*, 9(1), 1–30. https://doi.org/10.3390/en9010021.

Jelle, B. P., Breivik, C., & Drolsum Røkenes, H. (2012). Building integrated photovoltaic products: A state-of-the-art review and future research opportunities. *Solar Energy Materials and Solar Cells*, 100(7465), 69–96. https://doi.org/10.1016/j.solmat.2011.12.016.

Kirubakaran, D. D., Pitchaimuthu, S., Dhas, C. R., Selvaraj, P., Karazhanov, S. Z., & Sundaram, S. (2018). Jet-nebulizer-spray coated copper zinc tin sulphide film for low cost platinum-free electrocatalyst in solar cells. *Materials Letters*, 220. https://doi.org/10.1016/j.matlet.2018.02.122.

Kocer, C. (2019). *The Past, Present, and Future of the Vacuum Insulated Glazing Technology.* https://www.glassonweb.com/article/past-present-and-future-vacuum-insulated-glazing-technology.

Lee, T. D., & Ebong, A. U. (2017). A review of thin film solar cell technologies and challenges. *Renewable and Sustainable Energy Reviews*, 70, 1286–1297. https://doi.org/10.1016/j.rser.2016.12.028.

Luceño-Sánchez, J. A., Díez-Pascual, A. M., & Capilla, R. P. (2019). Materials for photovoltaics: State of art and recent developments. *International Journal of Molecular Sciences*, 20(4). https://doi.org/10.3390/ijms20040976.

Memon, S. (2013). *Design, Fabrication and Performance Analysis of Vacuum Glazing Units Fabricated with Low and High Temperature Hermetic Glass Edge Sealing Materials* [Thesis]. Loughborough University.

Memon, S., Farukh, F., Eames, P. C., & Silberschmidt, V. V. (2015). A new low-temperature hermetic composite edge seal for the fabrication of triple vacuum glazing. *Vacuum*, 120(Part A), 73–82. https://doi.org/10.1016/j.vacuum.2015.06.024.

Murugan, S., & Horák, B. (2016). A review of micro combined heat and power systems for residential applications. *Renewable and Sustainable Energy Reviews*, 64, 144–162. https://doi.org/10.1016/j.rser.2016.04.064.

Płaczek-Popko, E. (2017). Top PV market solar cells 2016. *Opto-Electronics Review*, 25(2), 55–64. https://doi.org/10.1016/j.opelre.2017.03.002.

Saifullah, M., Gwak, J., & Yun, J. H. (2016). Comprehensive review on material requirements, present status, and future prospects for building-integrated semi-transparent photovoltaics (BISTPV). *Journal of Materials Chemistry A*, 4(22), 8512–8540. https://doi.org/10.1039/c6ta01016d.

Selvaraj, P., Baig, H., Mallick, T. K., Siviter, J., Montecucco, A., Li, W., Paul, M., Sweet, T., Gao, M., Knox, A. R., & Sundaram, S. (2018). Solar energy materials and solar cells enhancing the efficiency of transparent dye-sensitized solar cells using concentrated light. *Solar Energy Materials and Solar Cells*, 175, 29–34. https://doi.org/10.1016/j.solmat.2017.10.006.

Selvaraj, P., Baig, H., Mallick, T. K., & Sundaram, S. (2018). Charge transfer mechanics in transparent dye-sensitised solar cells under low concentration. *Materials Letters*. https://doi.org/10.1016/j.matlet.2018.03.137.

Selvaraj, P., Ghosh, A., Mallick, T. K., & Sundaram, S. (2019). Investigation of semi-transparent dye-sensitized solar cells for fenestration integration. *Renewable Energy*, 141, 516–525. https://doi.org/10.1016/j.renene.2019.03.146.

Selvaraj, P., Roy, A., Ullah, H., Sujatha Devi, P., Tahir, A. A., Mallick, T. K., & Sundaram, S. (2019). Soft-template synthesis of high surface area mesoporous titanium dioxide for dye-sensitized solar cells. *International Journal of Energy Research*, 43 (1). https://doi.org/10.1002/er.4288.

Shukla, A. K., Sudhakar, K., & Baredar, P. (2016). A comprehensive review on design of building integrated photovoltaic system. *Energy and Buildings*, 128, 99–110. https://doi.org/10.1016/j.enbuild.2016.06.077.

Skandalos, N., & Karamanis, D. (2015). PV glazing technologies. *Renewable and Sustainable Energy Reviews*, 49, 306–322. https://doi.org/10.1016/j.rser.2015.04.145.

Sulaeman, U., & Zuhairi Abdullah, A. (2017). The way forward for the modification of dye-sensitized solar cell towards better power conversion efficiency. *Renewable and Sustainable Energy Reviews*, 74, 438–452. https://doi.org/10.1016/j.rser.2017.02.063.

Sun, Y., Shanks, K., Baig, H., Zhang, W., Hao, X., Li, Y., He, B., Wilson, R., Liu, H., Sundaram, S., Zhang, J., Xie, L., Mallick, T., & Wu, Y. (2018). Integrated CdTe PV glazing into windows: Energy and daylight performance for different architecture designs. *Applied Energy*, September. https://doi.org/10.1016/j.apenergy.2018.09.133.

Wang, M., Peng, J., Li, N., Yang, H., Wang, C., Li, X., & Lu, T. (2017). Comparison of energy performance between PV double skin facades and PV insulating glass units. *Applied Energy*, 194, 148–160. https://doi.org/10.1016/j.apenergy.2017.03.019.

Yan, J., & Saunders, B. R. (2014). Third-generation solar cells: A review and comparison of polymer: Fullerene, hybrid polymer and perovskite solar cells. *RSC Advances*, 4(82), 43286–43314). https://doi.org/10.1039/c4ra07064j.

10

HARVESTING THE SUN

Paving the way to a net zero carbon future

Luksa Kujovic

Experiencing cultures and environmental challenges

Experiencing diverse cultures has been possible for me due to living in six different countries and travelling to eleven others. Born in Kuala Lumpur, Malaysia, I was exposed to a rich blend of traditions, languages, and cultures, making for an exciting and unique childhood. Growing up in Malaysia was a feast for the senses, from the vivid colours of traditional clothing and festivals to the mouth-watering aromas of street food and hawker centres. However, amid this enchanting backdrop, I became intimately aware of a dark cloud that threatened the air I breathed – the 1997 Southeast Asian haze. This environmental calamity was the result of slash-and-burn farming techniques employed in Indonesia to clear land for agriculture (Ketterings, Wibowo, Noordwijk, & Penot, 1999). In terms of greenhouse gas (GHG) emissions, the 1997 Indonesia fires were the worst (GFED, 2015). Malaysia was one of the countries affected by the haze, which led to me developing respiratory health problems, which fortunately improved over time. This was my first encounter with the effects of climate change.

Just as I was about to start primary school, my family moved to the ancestral home of Sarajevo, Bosnia and Herzegovina. Sarajevo has a rich and complex history, influenced by Ottoman, Austro-Hungarian, and Yugoslavian rule. I was surrounded by different languages, traditions, and religions. However, the devastating effects of the Bosnian War in the 1990s left a profound mark on the city and its people. Sarajevo is recognised as one of Europe's most air-polluted cities due to high levels of fuel consumption, traffic, and inadequate air pollution reduction policies. The city experiences high levels of smog during winter due to the use of firewood and coal for heating and the geography of the surrounding mountains, which trap the

DOI: 10.4324/9781003380566-14

polluted air (IQAir, 2023). Despite the challenges, Sarajevo is a beautiful city with much to offer.

After a year in Sarajevo, it was time to relocate to Abu Dhabi, United Arab Emirates (UAE), followed by Dubai. While in the UAE, I observed that renewable energy sources were not widely utilised as the country relied heavily on its vast oil and natural gas reserves. Eight years later, my family moved to Doha, Qatar, where this was also evident. Living in the Middle East for over a decade was a unique experience that allowed me to learn about a diverse range of cultures. I then moved to Eindhoven, Netherlands, to pursue my undergraduate degree. It was refreshing to see renewable energy sources daily. I was particularly impressed to learn that all electric trains are fully powered by wind energy (*The Guardian*, 2017). My experiences in different countries made me reflect on the various environmental challenges and the need for sustainable solutions.

Finally, I moved to Loughborough, United Kingdom, where I became inspired to pursue a PhD to contribute to a better future. My experiences living in and visiting different countries have instilled in me a deep passion for sustainability, and I knew that I wanted to devote my life to this cause. I am excited to see where this journey takes me and to make a positive impact on the world.

A journey through my past research

I graduated from Eindhoven University of Technology with a BSc in Electrical Engineering. My BSc research project, which focused on "Maximising DER Penetration in Distribution Networks", was my first step towards contributing to a net zero future. The traditional unidirectional power flow from the transmission network (TN) to the distribution network (DN) has been transformed into a bidirectional flow (Miller, 2017). This is due to power injection by distributed energy resources (DERs) to the DN. DERs are small-scale power generation and storage technologies that can be located close to the point of use. They can include renewable energy sources such as photovoltaic (PV) panels, wind turbines, small hydroelectric generators, and non-renewable sources such as natural gas-fired generators. DERs can also include energy storage systems like batteries and thermal storage systems. DERs offer several benefits, including increased energy efficiency, reduced reliance on centralised power plants, improved grid resiliency, and reduced GHG emissions. They are essential to transitioning towards a more sustainable and decentralised energy system. My DER choice became very easy when I started to explore renewable energy systems; solar power stood out the most to me. The electricity generated from PV systems is as low as 2.4 US cents per kWh at the utility scale (LAZARD, 2023), which is significantly cheaper than conventional fossil fuels and most other renewable energy sources. Additionally, the GHG emissions from PV technologies are significantly lower than those from conventional fossil fuels, with thin film cadmium telluride (CdTe) modules having the lowest emission factors (Nelson, Gambhir, & Ekins-Daukes, 2014). This made solar power the obvious choice. The use of PV systems has been rising over the years (IEA, 2022) due to solar power's environmental and economic benefits. This increasing DER penetration has led to increased operational problems such as over/under-voltage and overloading, making it challenging for the transmission system operator (TSO) to call upon all available DERs. Flexibility services such as controlling reactive power and active loads are essential in maximising DER penetration. I used MATPOWER (Zimmerman, Murillo-Sánchez, & Thomas, 2011) to model a 31-bus DN and perform Newton AC power flow calculations in my research. The DN system had eight distributed generators, modelled as rooftop PV systems.

The load and PV generation data were acquired and scaled to fit the DN. The load for 100 houses was modelled using a bottom-up Markov Chain Monte Carlo approach (Nijhuis, Gibescu, & Cobben, 2016). The data used to model the PV generation were obtained from The Dutch PV portal system design model (Schepel et al., 2020). The system model includes real-time meteorological data from the Royal Netherlands Meteorological Institute (KNMI). Due to the stochastic behaviour of most DERs, it is hard to forecast how much power will be produced by these renewable energy sources. As a result, the uncertainty in the whole system is increased. When performing the power flow calculations, I took this uncertainty into account. I modelled the uncertainty using the Gaussian distribution, a continuous probability distribution type. I provided a method to call upon the DERs while ensuring that the DN operates within voltage constraints. This was done by implementing constraints for the operation of DERs. In case of voltage constraint violations, the distribution system operator (DSO) manages the load to maintain voltage limits. This allowed the DSO to utilise active loads as a flexibility service and maximise DER penetration while handling the increased uncertainty in the system. Maximising renewable energy penetration in existing distribution networks will make it easier to achieve a net zero carbon future. Through my BSc research project, I realised that integrating DERs, specifically PV systems, is crucial in making this transition possible.

Upon graduating from Loughborough University with an MSc in Renewable Energy Systems Technology, I knew that my next step towards contributing to a net zero future was to conduct research that could make a real difference. The MSc course was an excellent way for me to decide which renewable energy source to focus on, as it covered a variety of technologies, such as wind power, hydro power, solar power, and bioenergy. The idea of doing a PhD had never crossed my mind, but this changed when I started to deeply enjoy my MSc research project, titled "Degradation of Potential N-type Buffer Layers for Thin Film CdTe Solar Cells". It is well known that buffer layers significantly improve the performance of CdTe solar cells (Ferekides, Mamazza,

Balasubramanian, & Morel, 2005). A buffer layer is a thin layer of material placed between the CdTe absorber and the transparent conductive oxide (TCO) layer. The buffer layer plays an essential role in improving the performance and stability of the solar cell. The buffer layer helps optimise the interface between the CdTe absorber and TCO, improving the electrical properties and reducing interface recombination between these two layers. Buffer layers assist in achieving suitable band alignment between the absorber and TCO layer. This alignment is critical for efficient charge carrier transport and extraction. Using suitable buffer layers improves the cell's open-circuit voltage (Voc). The n-type buffer layer that was traditionally used in CdTe PV cells is cadmium sulphide (CdS), although this has largely been replaced in more recent devices. CdS buffer layers reduce solar cell performance due to absorption losses caused by the 2.42 eV band gap (Shafarman & Stolt, 2003). Furthermore, the band alignment at the CdS/CdTe interface is suboptimal (Song, Kanevce, & Sites, 2016). Researchers have studied higher band gap buffer layers such as magnesium zinc oxide (MZO), which has shown efficiency improvements due to decreased optical losses and improved band alignment (Munshi et al., 2018). The interface recombination was reduced due to the improved band alignment at the buffer layer/absorber interface. Unfortunately, MZO has been observed to degrade when exposed to atmospheric conditions (Bittau et al., 2019), decreasing device performance. This degradation is due to the magnesium oxide (MgO) reacting with atmospheric moisture to form magnesium hydroxide (Mg $(OH)_2$).

Studying alternative stable buffer layers is essential to ensure solar cells can achieve expected efficiencies over their lifetimes. Buffer layers which may be used instead of CdS include zinc oxide (ZnO), MZO, gallium-doped MZO (GMZO), cadmium zinc sulphide (CdZnS), and tin oxide (SnO_2). These materials have properties which allow them to work in PV devices. In my research, the environmental stability of ZnO, MZO, GMZO, SnO_2, and CdZnS was explored and compared in order to select the most suitable buffer layer material. I performed accelerated lifetime tests (ALTs) to determine their level of degradation when exposed to damp heat and ultraviolet (UV) radiation for 1,000 hours. I characterised the buffer layers based on their electrical, optical, chemical, and structural properties before and after the ALTs. The results showed that the samples with higher MgO content degraded more, which correlates with previous studies. I found that ZnO was the most environmentally stable buffer layer. The high transmittance and band gap of 3.25 eV indicate that ZnO is a suitable buffer layer for thin film CdTe solar cells. Due to the lower absorption losses, replacing CdS with ZnO is expected to improve the device's performance. Furthermore, replacing MZO with ZnO would allow devices to maintain the conversion efficiency for longer. My research highlighted the importance of studying alternative buffer layers due to the performance improvements they can provide in CdTe solar cells. It was a significant milestone in my academic journey and helped me realise the potential of renewable energy technology.

A look into my current research

During my PhD journey, I was fortunate enough to secure a three-month internship at First Solar, Inc., the global leader in thin film CdTe solar cells. It was an incredible opportunity to work alongside some of the brightest minds in the field, and I knew I would gain invaluable knowledge and experience during my time there. First Solar has achieved the record CdTe research cell efficiency of 22.1% (NREL, 2023). Before heading to First Solar, I had prepared a range of buffer layers to be used in their fabrication process. These buffer layers, including ZnO, CeZnO, SnO2 and SnZnO, were deposited using radio-frequency (RF) sputtering. Working with these layers had given me a solid understanding of their properties and performance but seeing them in action at First Solar was a whole new experience. During my internship, I witnessed and participated in First Solar's cutting-edge research and development efforts. I learned much about CdTe solar cells and how they could be optimised for maximum efficiency. One of the highlights of my time at First Solar was seeing the devices I had worked on come to life. I was thrilled that the devices incorporating a ZnO buffer layer achieved the best performance. We achieved a cell efficiency of 19.5% without an anti-reflective coating and optimisation, which was a significant accomplishment. While the devices utilising CeZnO and SnZnO did not reach comparable efficiencies, it was interesting to see what effect alloying had on device performance. Alloying with Ce and Sn improved the activation energy, indicating a reduction in interface recombination. This discovery emphasised the potential that these alloyed materials have.

As well as my time at First Solar, I also had the opportunity to visit the Next Generation Photovoltaics Centre at Colorado State University. There, I was able to fabricate CdTe devices for the first time. It was an eye-opening experience that made me appreciate how delicate and precise the fabrication process is. In order to achieve great results, researchers need to take extreme care at every step of the way and employ a range of characterisation techniques to fully understand the device's behaviour. My time at First Solar and Colorado State University allowed me to deepen my knowledge of CdTe solar cells and refine my research techniques. I am incredibly grateful for the opportunities and experiences I have gained during my PhD journey so far.

In my ongoing research journey, my primary focus is on optimising the buffer layers used in CdTe solar cells. I am investigating how chemical composition, thickness, substrate temperature, and oxygen levels impact the overall performance of the devices. To ensure the long-term efficiency and reliability of the cells, I am conducting accelerated lifetime testing, subjecting them to rigorous damp heat conditions for 1,000 hours. Developing suitable and stable buffer layers is crucial for reducing the cost of electricity generated by CdTe solar cells. This advancement is important in making renewable energy accessible to a broader population. My passion for this field is fuelled by the promise of providing clean and affordable energy to all.

After returning from First Solar, I eagerly resumed my work in the lab, focusing on the ongoing optimisation of the buffer layers. I conducted experiments involving the deposition of various ZnO thicknesses on three distinct substrates, allowing me to identify the optimal ZnO thickness and substrate. Additionally, I deposited a second batch of ZnO buffer layers at different substrate temperatures. These samples were then sent to First Solar for device fabrication. Following the characterisation of the devices, valuable insights were obtained regarding the optimal substrate, temperature, and buffer layer thickness. The thickness sweep led to a remarkable breakthrough, as a ZnO-based CdSeTe/CdTe device achieved an efficiency of 21.44%. Interestingly, this exceptional performance was achieved despite the buffer layer not being deposited at the optimal temperature. This means there is still room for improvement, and the new record for ZnO based devices is right around the corner.

My PhD project involves fruitful collaborations with world leaders in PV research, such as First Solar, National Renewable Energy Laboratory (NREL), Swansea University, and Colorado State University. This collaborative approach enables us to regularly exchange discoveries, allowing us to improve our understanding of thin film CdTe solar cells. Through our collective efforts, we will continue to see significant improvements in the performance of CdTe devices.

My future endeavours

It is important to continue improving the performance of thin film CdTe solar cells while maintaining low manufacturing costs. My plan is to continue with the buffer layer optimisation, and I will refine the thickness and temperature sweeps by incorporating finer increments around the identified "optimal" values. This granular approach will yield more precise and accurate values for the optimal parameters, enabling us to unlock even more significant potential. I am also committed to further exploring alloyed materials by optimising their chemical composition. Optimising the chemical composition of the CeZnO and SnZnO layers could improve the band alignment and reduce interface recombination. By optimising the buffer layer and ensuring environmental stability, I hope to replace existing buffer layers and further push the boundaries of CdTe devices.

As a PhD student, I have experienced the rollercoaster ride that this journey entails. It can sometimes feel like a dark tunnel, with the stress of experiments not going as planned, equipment malfunctioning, and the pressure to meet deadlines. However, with perseverance and dedication, I am starting to see the light at the end of the tunnel. One of the most significant aspects of pursuing a PhD is the numerous opportunities that await beyond graduation. While the future may appear uncertain, my aspiration to contribute to a brighter future remains resolute. Whether it leads me to the realms of academia or industry, I am committed to making a positive impact on the world.

In the landscape of challenges confronting humanity today, climate change looms as the biggest threat. Its far-reaching consequences demand urgent action to confront this global crisis and safeguard the prospects of future generations. It is easy to take our planet for granted, but we must cherish it and strive to preserve it for as long as possible. In this uncharted territory of academic exploration, I embrace the uncertainties and challenges that lie ahead. With every step, I am driven by the belief that the knowledge I acquire and my contributions will leave a lasting impact on society.

References

Bittau, F., Jagdale, S., Potamialis, C., Bowers, J. W., Walls, J. M., Munshi, A. H., … Sampath, W. S. (2019). Degradation of Mg-doped zinc oxide buffer layers in thin film CdTe solar cells. *Thin Solid Films*, 691, 137556.

Ferekides, C. S., Mamazza, R., Balasubramanian, U., & Morel, D. L. (2005). Transparent conductors and buffer layers for CdTe solar cells. *Thin Solid Films*, 224–229.

GFED. (2015). Regional highlights – Indonesia – emission estimates. Retrieved from Global Fire Emissions Database: https://www.globalfiredata.org/updates.html#2015_indonesia.

The Guardian. (2017). Dutch electric trains become 100% powered by wind energy. Retrieved from: https://www.theguardian.com/world/2017/jan/10/dutch-trains-100-percent-wind-powered-ns.

IEA. (2022). Renewables data explorer. Retrieved from IEA: https://www.iea.org/data-and-statistics/data-tools/renewables-data-explorer.

IQAir. (2023). Air quality in Sarajevo. Retrieved from IQAir: https://www.iqair.com/us/bosnia-herzegovina/federation-of-b-h/sarajevo.

Ketterings, Q. M., Wibowo, T. T., Noordwijk, M.v., & Penot, E. (1999). Farmers' perspectives on slash-and-burn as a land clearing method for small-scale rubber producers in Sepunggur, Jambi Province, Sumatra, Indonesia. *Forest Ecology and Management*, 157–169.

LAZARD. (2023). 2023 levelized cost of energy+. Retrieved from LAZARD: https://www.lazard.com/research-insights/2023-levelized-cost-of-energyplus/.

Miller, T. (2017). Bi-directional power flow: The new world order and what that means for the grid. Retrieved from IEEE Smart Grid: https://smartgrid.ieee.org/bulletins/november-2017/two-way-power-flow-the-new-world-order-and-what-that-means-for-the-grid.

Munshi, A. H., Kephart, J. M., Abbas, A., Shimpi, T. M., Barth, K. L., Walls, J. M., & Sampath, W. S. (2018). Polycrystalline CdTe photovoltaics with efficiency over 18% through improved absorber passivation and current collection. *Solar Energy Materials and Solar Cells*, 9–18.

Nelson, J., Gambhir, A., & Ekins-Daukes, N. (2014). *Solar Power for CO_2 Mitigation*. Grantham Briefing Papers.

Nijhuis, M., Gibescu, M., & Cobben, J. F. (2016). Bottom-up Markov Chain Monte Carlo approach for scenario based residential load modelling with publicly available data. *Energy and Buildings*, 121–129.

NREL. (2023). Best research-cell efficiency chart. Retrieved from https://www.nrel.gov/pv/cell-efficiency.html.

Schepel, V., Tozzi, A., Klement, M., Ziar, H., Isabella, O., & Zeman, M. (2020). The Dutch PV portal 2.0: An online photovoltaic performance modeling environment for the Netherlands. *Renewable Energy*, 175–186.

Shafarman, W. N., & Stolt, L. (2003). Cu(InGa)Se2 Solar Cells. In *Handbook of Photovoltaic Science and Engineering* (pp. 567–616). John Wiley & Sons Ltd.

Song, T., Kanevce, A., & Sites, J. R. (2016). Emitter/absorber interface of CdTe solar cells. *Journal of Applied Physics*, 233104.

Zimmerman, R. D., Murillo-Sánchez, C. E., & Thomas, R. J. (2011). MATPOWER: Steady-state operations, planning, and analysis tools for power systems research and education. *IEEE Transactions on Power Systems*, 12–19.

PART IV
TRANSPORTATION SECTOR

11

THE ROAD TO ELECTRIC VEHICLES

From architecture to portable battery storage via renewable energy research homes

Rebecca Roberts

Early research

From a young age I have always been interested in the built environment, specifically how a building or environment can make someone feel. Combining my love of art with my love of maths and science, architecture seemed like a natural progression for me to study at university. However, I was conscious that buildings are not always designed with nature in mind, so I decided to study a combined undergraduate master's degree in engineering and architecture – Architecture and Environmental Design at the University of Nottingham.

Within my first year of study, I realised that whilst designing buildings was a fascinating field, building services, and particularly renewable energy design, was far more exciting to me. In 2011 I started to consider the application of renewable technology, such as solar and heat pumps, in domestic properties, specifically retrofit solutions. Individually, these systems can be relatively easy to install if the appropriate systems have been implemented (for example, heat pumps require specific pipe diameters for radiators or underfloor heating). However, if not, the installation cost can result in a negative return on investment. Through this early research, I started looking at the Creative Energy Homes at the University of Nottingham. There are a variety of innovative homes on the site, and one was of interest as it had a combined heating/electricity system comprising a heat pump and solar photovoltaic array, along with various other low carbon interventions. This was, of course, a newly built property with the renewable technology installed at the time of build and combined with highly efficient insulation, innovative building design and a host of data monitoring. I wished to understand: (i) existing energy efficiency requirements in the UK for newly built homes and how they compare to the

DOI: 10.4324/9781003380566-16

state-of-the-art in terms of zero carbon footprint, (ii) the retrofit potential for existing housing stock in the UK, and (iii) the existing R&D of combined renewable generation systems, i.e., combined solar photovoltaics and thermal with heat pumps.

My studies at this point began to focus on the third point, the largest area of innovation – my argument being you cannot retrofit renewable generation technology to a home if we have not developed the technology sufficiently – and the property I focused on was home to an air source heat pump (ASHP). In 2011, these were relatively novel, and there was an enormous amount of research and opportunity to be explored, particularly around improving the coefficient of performance (COP) in low-temperature conditions.

Heat pumps are designed to take energy from their surroundings and dis-tribute it to a high-energy sink. Ground source is typically the most efficient due to the (usually) lower temperature difference between the source and sink, the downside being the need to bury large amounts of coil under-ground. ASHP is a non-invasive installation, taking heat from the air itself; however, efficiency is typically lower due to the more significant tempera-ture difference between the source and sink (Energy Saving Trust, 2022a; Gough, 2012). At the time of my research in 2011, there was concern around the COP for ASHPs as they had a minimum effective operational temperature of around 5°C, below which the efficiency dropped con-siderably, and additional support was required, e.g., gas boiler or other renewable technology such as a solar thermal collector (STC). STCs collect solar radiation and convert it into thermal energy. Solar thermal collectors are designed to maximise the absorption of solar heat and transfer it to a heat transfer fluid (usually a mixture of water and antifreeze). The heated fluid is then circulated through a loop to transfer the thermal energy to the heat pump unit (D'Antoni & Saro, 2012). Throughout 2010, the temperature reg-ularly dropped below this threshold, and therefore an ASHP alone would not be fit for purpose in an existing domestic property due to the insulation requirements (Gough, 2012). However, combined with highly thermally efficient wall insulation and STC, it would be possible for an ASHP to pro-vide most of the heating needs for newly built homes; retrofit to the existing housing stock in the UK would likely be unsuitable without significant insu-lation improvements (Gough, 2012; White et al., 2014). Twelve years after this research, ASHP efficiencies have drastically improved, operating at temperatures well below freezing with a relatively comfortable COP (Energy Saving Trust, 2022b). However, without the government grants currently available, retrofit is typically only affordable when replacing an end-of-life gas boiler. Otherwise, the payback period often exceeds the warranty period of the system, making retrofitting a challenge.

I graduated from the University of Nottingham in January 2012. Initially, I had planned on continuing for an additional year to complete my under-graduate masters, but the architecture side of my degree was not really of interest to me, and therefore I had a couple of options open to me. The first was a year out in industry at E.ON with part of their renewable energy research team. I was a runner-up in the UK Engineering Undergraduate of the Year Award and was offered a yearlong undergraduate placement. This would have meant I continued my undergraduate degree after that year but with some real-world commercial experience. The other option was provided by way of a renewable energy installation company I had been working with throughout the third year of my undergraduate studies. They had offered to sponsor me for an engineering doctorate with Loughborough University, as they were interested in the opportunities a combined heat pump with solar thermal collector technology could provide to new build properties. At the time, no one had investigated this in detail, so the scope for research, particularly from a PhD or engineering doctorate level, was intriguing.

I was very keen to be based in industry and to start work – student life had never really suited me – so I decided the context of research in a commercial setting through an engineering doctorate (EngD) would suit me. An EngD differs from a traditional PhD in three ways: (i) it is typically a four-year course rather than three; (ii) the first year is usually dedicated to undertaking a master's degree in a relevant subject; and (iii) due to the industry context, the thesis can be shorter and focused more on core research papers as opposed to a more traditional 50,000–80,000-word document.

Engineering doctorate

I therefore chose the engineering doctorate route and graduated in 2012 with a BSc in Architecture and Environmental Design. I then embarked upon an EngD with the Loughborough University Wolfson School of Engineering, concentrating on systems engineering. The first year was entirely focused on a master's in systems engineering, which covered the principles, focusing primarily on aerospace due to the university's strong ties with that industry. Initially, my research was focused on a combined energy system of a heat pump with solar generation due to the potential efficiency improvements a combined system could provide.

A combined heat pump with a solar thermal collector is a system that is designed to provide both space heating and domestic hot water (DHW) heating in a building. The system typically consists of two main components: a heat pump and a solar thermal collector (STC). Combining the solar thermal collector and the heat pump allows for renewable solar energy to be utilised as the primary heat source. The solar thermal collector provides thermal energy from the sun, while the heat pump maximises its efficiency by upgrading and distributing the collected heat throughout the building (Carbonell et al., 2014). The combined system offers the benefits of renewable energy utilisation, increased energy efficiency, and reduced reliance on traditional heating sources. It can be particularly advantageous in regions with ample solar radiation and a need for space heating and DHW heating in residential, commercial, or institutional buildings (Rad et al., 2013).

This can be enhanced further through deploying solar photovoltaic thermal (PVT) instead of solar thermal. PVT technology combines photovoltaic (PV) and thermal energy collection in a single solar panel. This hybrid system allows for the simultaneous generation of electricity and heat from sunlight. A solar PVT collector typically consists of PV cells, which convert sunlight directly into electricity, and a heat-absorbing element that extracts thermal energy from the sunlight. The PV cells capture solar radiation and convert it into electrical energy. At the same time the heat-absorbing element absorbs the excess heat from the PV cells or the sunlight that is not converted into electricity (Chandrasekar & Senthilkumar, 2021). The captured heat can be utilised for various applications, such as water heating, space heating, or even for generating steam in industrial processes. By extracting heat from the PV cells, solar PVT systems offer the advantage of cooling the PV cells, which can enhance their efficiency and lifespan (Ghadikolaei, 2021).

Through this early exploration, the challenge faced was the limited commercial potential of such a system, and it very quickly became clear that the commercial application would be purely for new builds as opposed to retrofit solutions (Lygnerud et al., 2021). At the same time, I had been increasingly interested in the opportunities of static battery storage, particularly combined with solar PV in a domestic setting. This took the application of energy storage I had explored during my undergraduate degree – in the form of heat – and applied it for power purposes. Static batteries were still extremely expensive and in relative infancy regarding the research opportunities for domestic applications; however, some business cases were forming for large-scale battery storage, such as those explored by Open Energi and Flexitricity (O'Dea, 2023). One benefit of static storage for domestic applications is that it can increase the usability of renewable generation, for example storing excess PV generation during the day to utilise during the evening, thereby reducing the payback period (Parra et al., 2014).

Electric vehicles (EVs) are becoming more popular, but in 2013, they were still very much in their infancy, with mileage ranges still being short compared to internal combustion engines (ICEs). The UK was putting in vast amounts of funding for both EV purchasing and charging infrastructure to promote the uptake of EVs across the UK and globally (*The Guardian*, 2011). Through my research on the applications of a combined home energy system, I was seeing more and more research around vehicle-to-grid.

The pivot to electric vehicles

Vehicle-to-grid (V2G) is a technological concept that enables the bidirectional flow of electricity between EVs and the power grid. In a V2G system,

EVs are not only seen as mobile energy consumers but also as potential energy sources and storage units. This technology allows EVs to draw power from the grid for charging and discharge electricity back into the grid during specific periods or under predetermined conditions. V2G leverages the flexibility and energy storage capacity of EVs to support grid stability, optimise energy management, and facilitate the integration of renewable energy sources. By utilising V2G capabilities, EVs can provide services such as peak shaving, load shifting, frequency regulation, and demand response, which help to enhance the power grid's overall efficiency, reliability, and sustainability (Gough, 2016). In 2013, the only hardware installed with V2G capability was in Japan, where the technology originated in the wake of the Fukushima disaster to provide power during an electrical blackout (Engerati, 2014; Haddadian, 2014).

During my early exploration into V2G, I also experienced a change in commercial sponsorship for my doctorate. This meant that I either had the option of continuing with my existing area of study – combined domestic renewable energy systems and securing a new sponsor for my research – or pivoting my research to the commercial opportunities and business cases for V2G. I chose to go down the V2G route as I felt the research potential and commercial applications were far superior to my previous study area.

Novelty of research: vehicle-to-grid

In early 2014, I started my research into V2G with Cenex, The Centre of Excellence for low carbon and fuel cell technologies, looking at the commercial applications of V2G. I specifically focused on those systems that could exist at large commercial premises or a disparate connection of multiple vehicles through a virtual power plant (VPP). A VPP is a network of decentralised power generation sources, energy storage systems, and demand response capabilities interconnected and coordinated through advanced digital technology (Pourghaderi et al., 2022). Unlike a traditional power plant, a VPP does not rely on a single physical location or centralised infrastructure but instead aggregates and manages distributed energy resources (DERs) that are geographically dispersed. The main objective of a virtual power plant is to maximise the efficiency, reliability, and flexibility of the overall energy system. By aggregating and orchestrating various distributed energy resources, a VPP can actively participate in energy markets, provide ancillary services to the grid, support grid stability, and enable the integration of renewable energy sources (Ullah et al., 2019).

We applied for two Innovate UK research grants, with both projects installing V2G units – the first in the UK – at Aston University and a domestic property in Loughborough, Leicestershire. My interest was less about the technology itself and more about the techno-economic feasibility of V2G, including the benefits it presented above static battery storage (Uddin et al., 2017).

Research up to this point had focused on example V2G scenarios of specific use cases and had not yet considered the broader applications of V2G and potential revenue opportunities they could provide (Gough, 2016). No software or analysis environment existed to assess the benefits from a commercial viewpoint. As such, I created a MATLAB model to evaluate the investment opportunity of V2G and local services case studies for future energy scenarios. The Feasibility Analysis Environment evaluated the economic benefit to both vehicle and building owners in installing V2G. The software could assess any case study with a collection of buildings, vehicles, PV or market demand, and energy scenarios were developed within the software to run case studies for economic evaluation, with these scenarios ranging from building peak shaving, tariff demand reduction, PV demand shifting, and energy market provision. By altering the number of vehicles being assessed, the software could also calculate infrastructure provision requirements and related capex costs (Gough, 2016; 2017).

Using Manchester Science Park as a case study, the software was evaluated to establish its usefulness in identifying energy support opportunities economically and technically viable to the case study. This was also supported by a verification and validation process to evaluate the software built against original stakeholder requirements. The results showed that providing energy to the capacity market with wholesale market trading was the most cost-effective option compared to the other six scenarios evaluated through the software. A net present value of over £420,000, including infrastructure costs, was calculated after a ten-year intervention using 50 electric vehicle batteries for one hour a day, three days a week, with an average energy discharge of 11.5kWh per day (Gough, 2016).

Start-up in the European Space Agency

This early-stage research was novel and led to several other Innovate UK funding calls in V2G and research projects for the team. I learnt a lot about bid writing during this time, particularly as a large amount of funding was available from various European funding bodies such as Climate-KIC, Interreg, and the European Space Agency.

At the end of 2017, I left Cenex to take up a new role at Kearney as Head of Solutions for a start-up called EV8 Technologies, becoming Chief Operating Officer in 2020. The organisation was founded primarily based on the work I had already been doing at Cenex, as we looked to commercialise the technology patents, IP, and learning that we had developed through numerous V2G research projects. The company was formally founded in 2018 with Kearney, Cenex, and Brixworth Technologies as shareholders. EV8 builds software solutions supporting EV adoption, integrating the electricity grid, and optimising energy. The company's vision is to enable the largest EV VPP platform globally, supporting both B2B and B2B2C customers in participating through an end-to-end suite of services. This is hugely exciting as we are creating an organisation able to commercialise on the learning and know-how we developed through various research projects. In 2018, we launched a project called Human Switch in parallel to the business launch. This project was funded by the UK Space Agency and administered by the European Space Agency (ESA), focusing on the development of two core software solutions – EV8 Switch and EV8 Live. Fortuitously, my time at Loughborough University studying the systems engineering master's as part of my EngD benefited this project, as ESA follows a rigorous systems engineering approach. By the end of this project I am confident that, should the need arise, our software is robust enough to be launched into space!

The solutions developed through Human Switch focused on two key areas – supporting the transition to EV (EV8 Switch) and generating additional revenue once EVs have been adopted (EV8 Live). EV8 Switch helps businesses and consumers make informed decisions about their transition from ICE to EV, simplifying the process using real-world driving and vehicle data. In September 2021, we launched the EV8 Switch app to the UK public through the Apple App and Google Play stores in partnership with NatWest Bank. Subsequently, we also launched corporate and fleet versions of the application. EV8 Live enables users to predict energy demand, and optimise the capital needed to spend on the supporting infrastructure without compromising how the vehicle is used, maximising cost savings and generating new revenues. By leveraging the insights from EV8 Switch, EV8 Live can help address various topics such as smart charging and V2G. EV8 Live will optimise vehicles' charging and discharging schedules whilst enabling them to continue to fulfil their primary requirements as a means of transport. Energy markets can be complex, and opportunities for V2G will vary from country to country; however, they can typically be categorised into three types: the capacity market, the energy market, and the balancing services market.

Through the Human Switch project, we developed a working prototype of the EV8 Live platform, during which we learnt a lot about the key elements that influence the success of the business case for V2G. The most crucial factor is the ability to predict the location and charging demand of EVs at a particular point in time (Andersen et al., 2018). We explored a few methods of doing this, from machine learning (Shipman et al., 2019) to recurrent neural networks (Shipman et al., 2021) and classifying driving behaviour to predict vehicle availability (Naylor et al., 2019).

Where high granularity of location data is unavailable, one method of increasing the predictability of charging demand, and therefore vehicle use, is classifying driver behaviours into archetypes. By grouping user driving patterns in a way that does not explicitly store large volumes of location data, it is possible to predict with some success the potential battery capacity of a vehicle at a given location (Naylor et al., 2019). An archetype of user behaviour categorises typical driving patterns into an "archetype", for example, a daily commuter. To create these categories, first, a dataset of driving behaviour was converted to a personal profile based on the statistical likelihood of vehicle use over a 24-hour period (Naylor et al., 2019). To create archetypes with a high level of accuracy, Dynamic Time Warping was used to calculate a match between a new user and established driving archetypes (Naylor et al., 2019). This, of course, is heavily reliant upon: a) the user understanding their driving behaviour, and b) the feed data being an accurate representation of typical driving, and is therefore open to inaccuracies, meaning this method is prone to a level of inaccuracy we wished to improve upon.

Humans are intrinsically predictable when looking at a large enough population; however, being able to predict the likelihood that an individual will perform a specific action, such as plugging a car into a charging point, is inherently more complex. To make EV8 Live as successful as possible, we needed to understand how humans behave when (a) asked to decide on making their vehicle available for V2G services and (b) their reliability in following through on their decision (Shipman et al., 2019). The critical element of understanding here is not if they do what they say they will but if we can accurately predict the outcome of their behaviour. For this research, the team at the University of Nottingham explored the benefits of a machine learning model to predict this behaviour, using a series of simplified simulations to replicate behaviour (Shipman et al., 2019). Machine learning was chosen for its ability to learn and adapt as more data is fed into the system, and results did show that it was possible to better predict the available capacity at a given time and the number of users that are needed to satisfy the required capacity (Shipman et al., 2019).

One thing that impacted this research was COVID-19, specifically the behavioural changes of people no longer allowed to leave their homes. As expected, the vehicles we monitored in 2019 have drastically different trip data to 2020. This causes any modelling to become irrelevant unless a model is developed that can take much smaller amounts of data to predict behaviour and, energy availability, thus reducing the prediction error. This behavioural change meant the model could be trained on radically different data sets, making it more robust. A deep recurrent neural network was coached using characteristic data of fleet behaviour and then tested using vehicle data from the pandemic with and without machine learning (Shipman et al., 2021). Machine learning significantly reduced prediction error, demonstrating the need to integrate this type of advanced prediction algorithm into the EV8 Live platform (Shipman et al., 2021). However, this was tested using a relatively small amount of data, demonstrating the need to expand the research using many more months of data to improve the robustness of prediction to make it suitable for the commercial application of EV8 Live (Shipman et al., 2021).

This behaviour modelling was conducted so as to be able to predict aggregated available energy capacity 24 hours ahead, based on data collected from the previous 24 hours for energy trading, the purpose of which is to trade on energy markets with the EV batteries managed by the EV8 Live platform. Early development of this model used time series forecasting, using historical data from a fleet of vehicles based at the University of Nottingham (Shipman et al., 2021). This was then extended to explore the ability of the model to adapt to simulated market events, such as increased reserve requirements. The research proved successful; however, further developments are required to simulate and then perform actual trading for various markets, such as capacity and reserve, and this will be the next step towards the commercial readiness of the EV8 Live service.

An alternative way of knowing when a vehicle will be connected to a charging device and how much capacity there will be for charging/discharging is to create an environment whereby the vehicle's driver must adhere to specific behaviour requirements. We are still in the early stages of this research, and therefore no definitive results have been produced. One key element is that limited deviation from a typical day-to-day pattern will have a much greater success rate than enforcing boundaries outside the limits of a vehicle's typical operation. All this research is being explored to develop the commercially ready version of EV8 Live by the end of 2023.

The future of energy storage

There is a concern that we need to do more to increase the uptake of EVs in the UK, in Europe, and globally, and a big part of this concern is around the infrastructure required (Department for Transport, 2022). Increasingly, drivers are naming affordability and lack of charging infrastructure as reasons for not getting an EV (World Economic Forum, 2021). We need to do lot more as a society to increase the amount of EV charging and address the concerns of price being a barrier to entry for adoption.

Increasingly, I am interested in how we can enhance the affordability of EVs, thus accelerating our 2030 targets for EV adoption. V2G directly ties into this, and one of the topics I have turned my attention to is the potential opportunity bundled service offerings could provide for EV uptake. This means a bundled service of vehicle, infrastructure, and energy model whereby the buyer or leaser of a vehicle agrees to operate the vehicle within certain boundaries or parameters, and as a result, the purchase or lease cost of the vehicle is lower. For example, vehicle and bidirectional charging infrastructure is provided to a fleet manager at a 10% discount; however, 40% of the vehicles must be connected to the bidirectional charging infra-structure between 4:00 pm and 5:00 am four days a week. Creating a solu-tion that would enhance the affordability of low carbon transport, or any low carbon or renewable technology, is, of course, beneficial to society and something that must be explored, so the opportunity to explore this further is extremely exciting.

References

Andersen, P. B.et al. (2018). *Added Value of Individual Flexibility Profiles of Electric Vehicle Users for Ancillary Services*. 2018 IEEE International Conference on Communications, Control, and Computing Technologies for Smart Grids. Aalborg, Denmark.

Carbonell, D., Haller, M. Y., & Frank, E. (2014). Potential benefit of combining heat pumps with solar thermal for heating and domestic hot water preparation. *Energy Procedia*, 57, 2656–2665.

Chandrasekar, M., & Senthilkumar, T. (2021). Five decades of evolution of solar photovoltaic thermal (PVT) technology – a critical insight on review articles. *Journal of Cleaner Production*, 322.

D'Antoni, M., & Saro, O. (2012). Massive solar-thermal collectors: A critical literature review. *Renewable and Sustainable Energy Reviews*, 16 (6), 3666–3679.

Department for Transport (2022). Taking charge: The electric vehicle infrastructure strategy. HM Government.

Energy Saving Trust (2022a). Air source heat pumps. [Online] Available at: https://energysavingtrust.org.uk/advice/air-source-heat-pumps/.

Energy Saving Trust (2022b). Air source heat pumps vs ground source heat pumps. [Online] Available at: https://energysavingtrust.org.uk/air-source-heat-pumps-vs-ground-source-heat-pumps/.

Engerati (2014). Nissan begins LEAF to home grid demand response testing. [Online] Available at: https://www.engerati.com/energy-retail/nissan-begins-leaf-to-home-grid-demand-response-testing/.

Ghadikolaei, S. S. C. (2021). Solar photovoltaic cells performance improvement by cooling technology: An overall review. *International Journal of Hydrogen Energy*, 46 (18), 10939–10972.

Gough, R. (2012). *Investigation into Low Carbon Heating Systems for the UK Domestic Sector*. [unpublished paper, Loughborough University].

Gough, R. (2016). *Electric Vehicle Energy Integration Scenarios: A Feasibility Analysis Environment*. [PhD thesis, Loughborough University]. Loughborough University Institutional Repository. https://repository.lboro.ac.uk/articles/thesis/Electric_vehicle_energy_integration_scenarios_a_feasibility_analysis_environment/9517361.

Gough, R., Speers, P., & Lejona, V. (2017). *Evaluating the Benefits of Vehicle-to-grid in a Domestic Scenario*. Electric Vehicle Symposium. Stuttgart, Germany, 9–11 October.

The Guardian (2011). UK government launches £5,000 electric car grant scheme. [Online] Available at: https://www.theguardian.com/environment/2011/jan/01/electric-car-grant-uk.

Haddadian, G. J. (2014). Power grid operation risk management: V2G deployment for sustainable development. Illinois Institute of Technology.

Lygnerud, K., Ottosson, J., & Kensb, J. (2021). Business models combining heat pumps and district heating in buildings generate cost and emission savings. *Energy*, 234.

Naylor, S., Pinchin, J., Gough, R., & Gillott, M. (2019). *Vehicle Availability Profiling from Diverse Data Sources*. 2019 IEEE International Conference on Pervasive Computing and Communications Workshops. Kyoto, Japan, 11–15 March.

O'Dea, S. (2023). Lithium-ion battery pack costs worldwide between 2011 and 2030. [Online] Available at: https://www.statista.com/statistics/883118/global-lithium-ion-battery-pack-costs/.

Parra, D., Walker, G. S., & Gillott, M. (2014). Modeling of PV generation, battery and hydrogen storage to investigate the benefits of energy storage for single dwelling. *Sustainable Cities and Society*, 10, 1–10.

Pourghaderi, N., Fotuhi-Firuzabad, M., Moeini-Aghtaie, M., & Kabirifar, M. (2022). Optimization model of a VPP to provide energy and reserve. In: *Scheduling and Operation of Virtual Power Plants*. Elsevier, pp. 59–109.

Rad, F. M., Fung, A. S., & Leong, W. H. (2013). Feasibility of combined solar thermal and ground source heat pump systems in cold climate, Canada. *Energy and Buildings*, 61, 224–232.

Shipman, R. *et al.* (2019). Learning capacity: Predicting user decisions for vehicle-to-grid services. *Energy Informatics*, 2 (37).

Shipman, R. *et al.* (2021). We got the power: Predicting available capacity for vehicle-to-grid services using a deep recurrent neural network. *Energy*, 221.

Shipman, R. *et al.* (2021). Online machine learning of available capacity for vehicle-to-grid services during the coronavirus pandemic. *Energies*, 14 (21).

Uddin, K. *et al.* (2017). Techno-economic analysis of the viability of residential photovoltaic systems using lithium-ion batteries for energy storage in the United Kingdom. *Applied Energy*, 206, 12–21.

Ullah, Z., Mokryani, G., Campean, F., & Hu, Y. F. (2019). Comprehensive review of VPPs planning, operation and scheduling considering the uncertainties related to renewable energy sources. *IET Energy Systems Integration*, 1 (3), 147–157.

White, J., Gillott, M., & Gough, R. (2014). Investigation of a combined air source heat pump and solar thermal heating system within a low energy research home. In: *Progress in Sustainable Energy Technologies: Generating Renewable Energy*. Springer, pp. 355–368.

World Economic Forum (2021). 4 reasons why electric cars haven't taken off yet. [Online] Available at: https://www.weforum.org/agenda/2021/07/electric-cars-batteries-fossil-fuel/.

12

AFRICA'S TRANSITION TOWARDS NET ZERO IN THE TRANSPORT INDUSTRY

Richie Moalosi, Yaone Rapitsenyane and Oanthata Jester Sealetsa

The spark towards sustainable transportation

The authors are industrial design lecturers in the Department of Industrial Design and Technology at the University of Botswana, Botswana. We developed a common interest in design for sustainability and formed an informal group where we share experiences and publish together. In 2022, Botswana's Ministry of Tertiary Education, Science, Research and Technology approached a number of individuals and formed a team to examine the modalities of developing electric vehicles for Botswana. This was against a background that the world is moving towards sustainable transport to help combat climate change. This sparked our interest in deeply understanding the status of sustainable transportation in Africa; hence this study.

Transition to electric vehicles

Climate change affects Africa in various ways, for example through prolonged droughts and perennial floods, as well as exacerbating problems directly resulting from human activity such as high pollution rates, environmental degradation, reduction of forests, etc. Despite these environmental challenges, Africa is the fastest urbanising region in the world due to rapid population growth, which strains existing infrastructure, transportation, and energy (Dioha et al., 2022). In most cases, urbanisation comes with motorisation. This has been accelerated by a massive increase in imported second-hand light-duty vehicles from other regions (Gorham, 2017). Assuming this culture of importation of second-hand light-duty internal combustion engine (ICE) vehicles is sustained, the importation of second-hand electric vehicles is likely to increase carbon dioxide (CO_2) emissions since handling battery

DOI: 10.4324/9781003380566-17

pack waste will probably become a challenge in Africa. Recycling electric vehicle (EV) battery packs will need to be a systemic part of the EV industry to facilitate the recovery and re-use of lithium from lithium-ion batteries. Carbon dioxide emissions from the transport sector are increasing at a rate of 7% per annum in Africa, which is considerably faster than other continents (Gorham, 2017). Gorham further states that the proportion of carbon dioxide emissions generated from the transport sector is higher in Africa than in other regions worldwide. For example, the average rate of vehicle pollution in the USA is 0.8%, and 0.12% in the UK (Ayetor, Mbonigaba, Ampofo, & Sunnu, 2021). On a per capita basis, the transport sector's carbon dioxide emissions are growing faster than any other source of energy-related carbon dioxide emissions across the continent. This challenge is accelerated by a lack of vehicle maintenance, an ageing vehicle fleet, poor fuel quality, the absence of vehicle standards, and poor vehicle standard enforcement (Ayetor, Mbonigaba, Ampofo, & Sunnu, 2021).

Rising oil prices and environmental crises such as global warming have contributed to the spread of electric vehicles (Amedokpo & Boutueil, 2022). Simultaneously there is a need to reduce the greenhouse gas emissions from vehicles in Africa. The electrification of the vehicle fleet provides an opportunity to reduce the carbon footprint resulting from vehicle emissions. EVs present an excellent prospect for decarbonisation (Ayetor, Mbonigaba, Ampofo, & Sunnu, 2021). As the world transitions to electric mobility, African countries will experience higher per capita CO_2 emissions, with continued use of petroleum and no explicit targets for their transition to electric vehicles. African countries can skip the fossil fuel-based revolution of the transport sector to offset CO_2 emissions. Leapfrogging fossil-fuelled transportation promises environmental, economic, human health, and infrastructure benefits for the continent. EVs are vital to developing a sustainable and low-carbon future for Africa.

EVs can be harnessed to improve Africa's energy security, achieve sustainable development, meet the requirements of various sustainable development goals, and increase a nation's attractiveness for investment (Xu et al., 2021). Glitman et al. (2019) state that EVs are three times more efficient than ICE vehicles. EVs also present a vital prospect to improve air quality and reduce noise pollution. Economically, EVs can promote public transport investment, making public transport more efficient, attractive, and affordable. While some African governments have taken positive steps to reduce fossil fuel use, such efforts must be scaled up and replicated where they have proven successful. African governments should develop enabling policy frameworks and integrated transport planning, make EVs affordable, generate reliable electricity, raise EV awareness, and support infrastructure in urban and rural areas. This chapter is divided into four sections. In the first section, four types of EVs are discussed. The second section focuses on case studies of electric mobility in Africa. The third section deals with challenges associated with acquiring EVs in Africa. The fourth section discusses the potential for switching to EVs.

Types of electric vehicles

There are four types of EVs. Battery electric vehicles (BEV) run entirely using an electric motor and battery without using a traditional ICE in any way. The EV must be plugged into an external source to recharge the battery. BEVs can recharge batteries through regenerative braking, which uses the vehicle's electric motor to slow it down whilst recovering some energy which would otherwise be converted to heat by the brakes (Ramya et al., 2021).

Hybrid electric vehicles (HEV) have two matching drive systems: a fossil-fuel driven ICE and an electric motor with a battery. The engine and electric motor can simultaneously turn the transmission, and the transmission then turns the wheels. HEVs cannot be recharged from the electricity grid, although they can recover some energy through regenerative braking. However, all their energy comes from gasoline. They can deliver fuel savings of up to 39% (Ayetor, Mbonigaba, Sunnu, & Nyantekyi-Kwakye, 2021).

Plug-in hybrid electric vehicles (PHEV) use an electric motor and battery that can be plugged into the power grid to charge the battery, but also have the support of an ICE that may be used to recharge the vehicle's battery and/or to replace the electric motor when the battery is low (Ramya et al., 2022. Since PHEVs use electricity from the power grid, they regularly realise more savings in fuel costs than traditional hybrid electric vehicles. A PHEV has a modest rechargeable battery, which delivers fuel savings of up to 60% (Goel et al., 2021).

Fuel cell electric vehicles (FCEV) produce electricity from hydrogen or another similar fuel using fuel cells rather than drawing electricity from a battery. FCEVs do use a battery for recapturing braking energy, however, providing extra power during short acceleration events, and smoothing out the power distribution from the fuel cell, with the option to turn off the fuel cell when power requirements are low. The size of the hydrogen fuel tank determines the quantity of energy stored.

Electric mobility case studies in selected African countries

Different African countries have embraced EVs, though the uptake still needs to be improved. For example, Ayetor, Mbonigaba, Ampofo, and Sunnu (2021) report that in 2020, Ghana had 5,693 EVs (94% PHEV and 4% HEV) representing 0.4% of the vehicle fleet. In 2020, Mauritius had 16,109 HEV, PHEV, and 331 BEV, representing 3.6% of the vehicle fleet (NLTA, 2020). Moreover, in 2020, Nigeria imported 60,773 EVs (24,926 HEV and 35,847 PHEV), representing 0.5% of the vehicle fleet (ITC, 2020). As of 2019, South Africa had 4,998 EVs composed of 3,879 HEV, 574 PHEV, and 545 BEV, representing 0.04% of the vehicle fleet (Ayetor, Mbonigaba, Ampofo, & Sunnu, 2021). Out of 2.2 million registered vehicles in Kenya, there are an estimated 350 EVs, representing 0.01% of EVs. The Kenyan government aims to increase EV ownership to 5% of all registered vehicles by 2025. The above statistics indicate that the HEV and PHEV dominate the African EV market. These figures are low, but these countries have the highest EV adoption rates in Africa. However, the promotion of renewable energy, the use of 4IR disruptive technologies, and the youth population provide prospects to boost EV uptake in Africa. Africa has a tremendous renewable energy mix, which will help to lower emissions and raise the chances of achieving net zero.

In Rwanda, the dominant mode of public transport is motorcycles (moto-taxis) that pollute the air, contribute to climate change, keep demand for imported fuel high, and contribute significant greenhouse gas emissions. In 2019, Rwanda launched e-motos with the intent to expand across Africa. This is bringing about a mass-market transition to electric mobility, which is cost-effective to buy and maintain in Africa. Rwanda aims to ensure that half of the motorcycles in Africa are electric by 2030. Rwanda sells brand-name electric motorcycles for a lower price and converts popular ICE versions into electric models by retrofitting them with batteries. This reduces emissions by 75%, ushering in a cleaner, more cost-effective change towards net zero.

Uganda hopes to achieve a technological leap in the automotive industry by developing a local EV industry (Amedokpo & Boutueil, 2022; Madanda et al., 2013). Uganda's state-owned nascent car manufacturer developed a 9-metre solar-powered electric city bus. The bus is operated from lithium-ion batteries that power an electric motor coupled to a 2-speed pneumatic shift transmission. The bus has a capacity of 90 passengers. The electric motor delivers up to 245 kW. The bus's floor is bamboo, while local resources were used to build the body and structure.

The Kenyan government has rolled out a series of policies to encourage EV use and production. Local companies have teamed up with international consortiums to produce EVs. For example, a start-up, BasiGo, developed electric passenger buses in Kenya. The buses can travel 250 kilometres on one charge. These buses are assembled in Kenya using parts designed by Chinese car-making giant BYD. Similarly, Roam Motors, a consortium between Swedish–Kenyan companies, has also begun piloting Nairobi electric buses. Roam Motors is to embark on mass-producing electric buses and motorcycles. These companies are expanding their operations to other African countries, such as South Africa, Egypt, Tunisia, and Mauritius.

South Africa is viewed as the leader in EV production. However, only a few manufacturers, including Toyota, Mercedes-Benz, and BMW, produce hybrids classified as new energy vehicles (NEVs). The South African government has promised an EV roadmap to speed up sales and production, but it is progressing at a snail's pace. In 2021, the government published the Green Paper on Advancement of NEVs after massive industry consultations. The Green Paper deals with the support and infrastructure investment levels needed to promote EV uptake. It examines the financing and tax system needed to create a resilient raw material supply network to assist South African efforts to be a global player in NEV manufacturing. The Green Paper also outlines how South Africa must maintain preferential entry to prominent trading associates to maintain global competitiveness and foster innovation.

The Ministry of Energy in Ghana reports that more than 1,000 electric vehicles are registered nationwide. Ghana aims to achieve carbon neutrality by 2050 by reducing CO_2 emissions by introducing EVs. However, Onai and Ojo (2017) and Wahab and Jiang (2019) argue that research must be conducted to understand the factors influencing consumers' EV purchase intentions. Ackaah, Kanton, and Osei (2021) outline some barriers inhibiting EV adoption in Ghana, such as frequent power outages, rapid battery degradation, high initial cost, and increased stresses on power system equipment prohibiting the massive adoption of EVs. The authors advise adding more battery switch stations to ensure an uninterrupted power supply for EVs. The private sector is assisting Ghana in developing green mobility. For example, Total Energies has installed an EV charging station in Accra – Ghana's capital city – and aims to roll out this initiative to other cities. Furthermore, Solar Taxi – a local company that assembles, sells, and leases electric motorbikes and sells imported EVs – is planning to start assembling EVs locally. All these efforts promote EVs in Ghana and marshal green and sustainable technology for the future.

These case studies show that the development and production of electric mobility in Africa has only just started. However, advancing the green agenda in electric mobility is facing challenges that African governments need to address. The following section discusses the barriers that hinder EV uptake in Africa.

Challenges and opportunities on the uptake of EVs

Several barriers have contributed to the slow development and uptake of electric mobility in Africa. These include access to electricity, electric vehicle national policy, development of EV standards, EV charging technology and infrastructure, African tax regimes, environmental (climate change) education, EV capacity building, and manufacturing prospects.

Access to electricity

This is a significant impediment to EV access in Africa. Ayetor, Mbonigaba, Ampofo, and Sunnu (2021) reported that in 2019, access to electricity in Sub-Saharan Africa was 48% compared to a world average of 90%. The United Nations Environment Programme state that electricity generation capacity and access remain impediments across the continent. It is estimated that over 640 million Africans have no access to electricity. In some cases, where there is access, the electricity supply is unreliable. This explains why most electric vehicles imported into Africa are HEVs and PHEVs. African governments should ensure EVs' maintenance and service centres and a stable and reliable green power supply.

Renewable energy development, e.g., solar energy, will assist in powering EVs with sustainable, clean energy by setting up charging stations. Africa is a perfect example of how e-mobility can impact climate change. However, Africa should move away from energy generated from fossil sources, e.g., coal, because such a national energy mix makes it hard to attain net zero. Charging EVs from the grid will produce a rebound effect where energy demand increases, increasing CO_2 emissions from coal-generated electricity. The deployment of EVs in Africa would be cost-effective using renewable energy-powered charging sources and/or smart charging stations to offset electricity generation deficits. EV charging powered by distributed and decentralised renewables has the potential to keep the systems clean and more reliable with less load. In this regard, more research is needed to assist Africa in effectively implementing smart charging systems.

Though there are challenges to accessing and supplying electricity, African cities have access to electricity mainly generated through thermal power using fossil sources, which results in carbon emissions. This provides an opportunity to deploy EVs to reduce the carbon footprint, provided charging stations are built in cities and rural areas.

Electric vehicles national policy

African countries still need to develop an EV national policy. The policy must provide a roadmap to what needs to be achieved, target timelines, increase supply, set vital uptake targets, and stimulate demand (Lynskey et al., 2022). Where EV policies are in place, faster implementation is constrained by a lack of funding. The policy could assist in educating users on the advantages of EVs, and promotional measures, e.g., tax incentives for EV owners, public EV charging infrastructure development, etc. The proposal of a staged and localisation-focused manufacturing roadmap, considering potential benefits for the nation's manufacturing ecosystem, might increase electric mobility and stimulate the development of electric vehicles. The policy should guide phased manufacturing programmes for EV components and batteries, transport systems, standards for installation of charging infrastructure, payment systems, sustainable business models, rebates and incentives for users, manufacturers, and transport service providers. Lynskey et al. (2022) and Begley et al. (2016) argue that a comprehensive policy is needed to promote the transport fleet's decarbonisation. A fast and smooth transition for individuals and businesses coupled with coordinated and sustained policy support to help the EV market transition through its emergent stages into a well-established marketplace are needed. The African EV policies should set ambitious EV sales targets to provide certainty for the market and gain additional savings, which will benefit consumers and the climate. The policy could also assist African countries in declaring EV electrification targets, e.g., to have achieved 100% zero emissions by a set year. The current lack of EV national policies in Africa does provide an opportunity for researchers to assist countries in developing these policies. Well-developed and funded policies will guide EV development and transition to a circular economy.

Development of EV standards

From the reviewed case studies, each country developing EVs has ambitions to export. However, most countries have no defined EV standards. The national standard bodies need to develop standards for EVs to safeguard user health and safety, and environmental sustainability (Ayetor, Mbonigaba, Ampofo, and Sunnu, 2021). The regulating framework should include battery charging systems and infrastructure, disposal and management of batteries, vocational standards, homologation, and health and safety (Ayetor, Mbonigaba, Ampofo, and Sunnu, 2021). Das et al. (2020) argue that EV charging standards relate to components, grid integration, and safety standards. The national standards can domesticate the UNECE R100 standard that approves new electric vehicles to pass homologation and tests of components against electrical shocks and fire. The International Electro-Technical Commission and Society of Automotive Engineers standards cover the charging infrastructure and components. The Underwriters Laboratories and Institute of Electrical and Electronics Engineers produce the grid integration standards. Unfortunately, most African countries do not have EV standards to safeguard the health and safety of users and the environment. Once again, this is another opportunity for researchers to assist African countries in developing EV standards.

EV charging technology and infrastructure

One main impediment that suppresses electric vehicle adoption rates in Africa involves recharging station availability. The charging technology and infrastructure throughout Africa are insufficient in quantity and quality, resulting in fuel range anxiety (Moeletsi, 2021; Bonges & Lusk, 2016). Only a few countries, such as South Africa, Mauritius, Kenya, and Uganda, have limited charging stations. The existing power networks must be improved for an increased EV use scenario as the market begins to push EVs into the African market. Even if charging points are accessible along pre-planned routes, charging must be fast. For example, if a single car takes about eight hours to recharge (in the case of a home, overnight or workplace parking lot charger), this is inefficient and acts as a barrier towards EV adoption. African countries should invest in efficient and fast charging networks. EV users living in family houses with garages and workplaces should be incentivised to install photovoltaic charging points for easy access from public charging points (Buresh et al., 2020). African countries should commit financial resources to installing new charging points, upgrading power grids, and increasing renewable energy generation capacity.

African tax regimes

Tax regimes in Africa prove to be a significant stumbling block to reducing the EV purchase price. Most countries impose import duties as high as 40%. This discourages the import of EVs and thereby contributes to high carbon emissions. Ayetor, Mbonigaba, Ampofo, and Sunnu (2021) report that Mauritius and Egypt are the only countries with reduced or waived import duties. For example, Mauritius waives excise duty and reduces registration duty, import duty, and road tax on imported EVs (EVCONSULT, 2020). Egypt provides 100% exemptions from import duties on EVs, excluding buses. African countries must review the tax regimes as part of their EV incentive programmes to encourage a shift towards sustainable transport mobility.

Environmental (climate change) education

In a study conducted in Ghana, Ayetor, Mbonigaba, Ampofo, and Sunnu (2021) report that emissions reduction was not one of the essential attributes regarding the purchase intention of EVs. Consumers were more concerned about the driving range and available EV infrastructure. Moeletsi (2021) reports that EV technology is considered disruptive, and its uptake is a challenge because it depends on consumers' preparedness and the modification of their behaviour to embrace new technology. Public education is vital to ensure widespread understanding of why there is a need to transition from fossil fuel-powered vehicles and energy sources to renewable energy to address climate change challenges. Provision and accessibility of charging stations and establishing exclusive metropolitan bus rapid transit systems modelled around EV buses may increase consumer confidence in owning and using EVs and facilitate sustainable development in Africa. For passenger cars, public confidence in Africa in the ability of EVs to adapt to underdeveloped road infrastructure and lifestyles of city-to-rural and rural-to-city settings is essential.

EV capacity building

EVs in Africa are a new phenomenon. Most countries still need to build a skill base for both imported EVs and those developed locally. Therefore, EV maintenance, research, and training have a broad skills gap. Ayetor, Mbonigaba, Ampofo, and Sunnu (2021) state that skilled EV maintenance technicians, planners, managers, and spare parts are rare and sometimes non-existent in Africa. Re-tooling of professionals in the automotive industry is a good starting

point for developing capacity in the EV space. There is a need for close cooperation and collaboration between African governments, industries, universities, and research institutions. This tripartite partnership can play a critical
role in building capacity and responding to the needs of the EV industry. Wood
& Mattson (2016) identify the lack of contextual knowledge as the leading
cause of the failure of most designs in developing countries due to cultural and
societal perceptions. The involvement of African researchers is more critical to
exploring the electrification of mobility based on context-relevant literature
stemming from a need to indigenise the technical aspects of EVs as well as the
social, cultural, economic, and political values that are specific to the continent
(Amedokpo & Boutueil, 2022; Collett et al., 2021). Research is needed to
address the impact of EVs on the national grid to establish a long-term implementation strategy, build local knowledge about EVs, and adapt their designs
(Amedokpo & Boutueil, 2022; Koberstaedt et al., 2018). The industry needs its
fast uptake of existing technologies so that what works best from research and
industry can be harmonised and adopted at a policy level.

Manufacturing prospect

Addressing the skills gap will be a massive prospect for manufacturing African
EVs. Manufacturing EVs is less complicated and presents an excellent opportunity for African countries to domesticate the production processes (Brönner et
al., 2019). This will assist in overcoming the EV purchase price barrier, which
remains an obstacle for Africa to transit to net zero. Africa is gifted with natural
resources in the value chain essential to manufacture EVs. Lithium-ion batteries
will be critical to closing the EV energy gap. For example, Zimbabwe produces
lithium, the Democratic Republic of Congo and Senegal produce cobalt, and
South Africa and Botswana produce nickel and manganese. African countries
have a substantial prospect to manufacture EV batteries which can be used and
exported globally. Manufacturing and trade agreements must be enacted
between African countries to realise such endeavours.

Meeting the growing battery demand at a lower economic, environmental,
and human cost will require high-quality recycling, upcycling, and repur-
posing, which must be improved across most African countries. Research on
the safe disposal of e-waste, recycling, and upcycling of used batteries will
be needed for EV maintenance. Such an approach is sustainable because it
prevents the extraction of new raw materials, which has the potential to
damage the environment.

Scope for further research

There is potential for African countries to leapfrog using EVs as their penetration is still shallow globally. The automotive industry in Africa is also still
underdeveloped, presenting an opportunity to set up state-of-the-art EV

manufacturing and assembly plants with no costs needed to retrofit existing facilities. This provides an opportunity to orient the vehicle fleet profile to EVs and create rapid change. Due to the low access to electricity, hybrid vehicles offer an opportunity for some African countries to move towards the electrification of their vehicle fleet. Lack of access to and supply of electricity in most African countries is a significant barrier to EV deployment. The importation and growth of local hybrid EVs in Africa present more than just an opportunity to clean air and improve sustainable transportation. They enable local manufacturing industries to create new jobs in the green sector. African governments should increasingly target this growth and deliberately attract investment. If EVs are produced in Africa, this will provide benefits such as job creation, new skills and technology development, and EV manufacturing industry upgrades.

The reviewed case studies across Africa, though at the infancy stage, show signs of an emerging green market for EV mobility. The private sector in most African countries spearheads most EV initiatives. African governments must invest in this area by creating a conducive environment through development of policies and regulations that conform to international EV standards.

Moreover, Africa must move towards using renewable energy sources. The transition to renewable energy sources for electricity generation should be prioritised in Africa as part of holistic leapfrogging. Significant increases in renewable energy in the national energy mix are needed to power EVs and reduce the carbon footprint. The opportunities are open for companies developing EVs and researchers who can guide the development of national electric mobility policies and indigenous electrification initiatives and develop capacity-building programmes for EVs across Africa.

Africa has the potential to leap to EV technologies by providing infrastructure needs, promoting investment opportunities, and financing renewable energy technologies, using sustainable energy sources, developing conducive policy frameworks and opportunities that promote local manufacturing of spare parts and maintenance, developing incentive-based regulatory schemes, building local technical capacity, and scaling up e-mobility pilots taking place in various countries. Undoubtedly, such EV initiatives lessen air pollution, accelerate progress towards Sustainable Development Goals, especially climate goals, create green jobs, and help Africa move towards a circular economy. The challenges reviewed offer PhD researchers an opportunity to contribute to Africa's a green economy.

References

Ackaah, W., Kanton, A. T., & Osei, K. K. (2021). Factors influencing consumers' intentions to purchase electric vehicles in Ghana. *Transportation Letters*, 14, 1031–1042.

Amedokpo, Y. T., & Boutueil, V. (2022). What place for electric vehicles as a research object and a practical alternative to internal combustion engine vehicles in Africa? Toward a research agenda based on a systematic literature review and a census of electromobility projects. *Transportation Research Record: Journal of the Transportation Research Board*. https://doi.org/10.1177/03611981221116355.

Ayetor, G., Mbonigaba, I., Ampofo, J., & Sunnu, A. (2021). Investigating the state of road vehicle emissions in Africa: A case study of Ghana and Rwanda. *Transportation Research Interdisciplinary Perspectives*, 11, 100409.

Ayetor, G. K., Mbonigaba, I., Sunnu, A. K., & Nyantekyi-Kwakye, B. (2021). Impact of replacing ICE bus fleet with electric bus fleet in Africa: A lifetime assessment. *Energy*, 221, 119852. doi:10.1016/j.energy.2021.119852.

Begley, J., Berkeley, N., Donnelly, T., & Jarvis, D. (2016). National policymaking and the promotion of electric vehicles. *International Journal of Automotive Technology and Management*, 16 (3), 319–340, https://dx.doi.org/10.1504/IJATM.2016.10001648.

Bonges, H. A., & Lusk, A. C. (2016). Addressing electric vehicle (EV) sales and range anxiety through parking layout, policy, and regulation. *Transportation Research Part A: Policy and Practice*, 83, 63–73.

Brönner, M., Ampofo, J., Fries, D., & Lienkamp, M. (2019). Configuration parameters within electric vehicle production strategies in sub-Saharan Africa – the car mobility case. *Procedia CIRP*, 86, 288–293.

Buresh, K. M., Apperley, M. D., & Booysen, M. J. (2020). Three shades of green: Perspectives on at-work charging of electric vehicles using photovoltaic carports. *Energy Sustainable Development*, 57, 132–140. doi:10.1016/J.ESD.2020.05.007.

Collett, K. A., Hirmer, S. A., Dalkmann, H., Crozier, C., Mulugetta, Y., & McCulloch, M. D. (2021). Can electric vehicles be good for sub-Saharan Africa? *Energy Strategy Review*, 38, 100722. doi:10.1016/J.ESR.2021.100722.

Das, H., Rahman, M., Li, S., & Tan, C. (2020). Electric vehicles standards, charging infrastructure, and impact on grid integration: a technological review. *Renewable Sustainable Energy Review*, 120, 109618.

Dioha, M. O., Duan, L., Ruggles, T. H., Bellocchi, S., & Caldeira, K. (2022). Exploring the role of electric vehicles in Africa's energy transition: A Nigerian case study. *iScience*, 25 (3), 103926. doi:10.1016/j.isci.2022.103926.

EVCONSULT (2020). A 10-year electric vehicle integration roadmap for Mauritius. https://www.google.com/url?sa=t&rct=j&q=&esrc=s&source=web&cd=&ved=2ahUKEwi70Y j_wKbyAhVfAWMBHXtFALcQFnoECAIQAQ&url=https%3A%2F%2Fpublicutilities.go vmu.org%2FDocuments%2F2020%2FReports%2526Publications%2FElectric%2520 Vehicle%2520Integration%2520Roadmap.pdf&usg=AOvVaw0V6SLkHhU5RfpPdClc SZ9C. Accessed 24 November 2022.

Glitman, K., Farnsworth, D., & Hildermeier, J. (2019). The role of electric vehicles in a decarbonised economy: Supporting a reliable, affordable, and efficient electric system. *Electronics Journal*, 32 (7), 106623.

Goel, S., Sharma, R., & Rathore, A. K. (2021). A review on barriers and challenges of electric vehicles in India and vehicle to grid optimisation. *Journal of Transportation Engineering*, 4, 100057.

Gorham, R. B. (2017). Prospects for decarbonising transport in Africa. *Energy and Transportation in the Atlantic Basin*, 127–149.

ITC (2020). Motor cars and other motor vehicles principally designed for the transport of persons. https://www.trademap.org/Country. Accessed 24 November 2022.

Koberstaedt, S., Kalt, S., Fürst, L., Lin, X., & Lienkamp, M. (2018). *Definition of Requirements for a New Vehicle Concept for Sub-Saharan Africa – Load Collectives for Battery and Electric Motor*. Proceedings of the 21st International Conference on Electrical Machines and Systems (ICEMS), Jeju, South Korea.

Lynskey, R., Rowe, H., & Ashaq, M. (2022). Accelerating EV uptake: Policies to realise Australia's electric vehicle potential. https://www.climateworkscentre.org/wp-content/uploads/2022/08/Accelerating-EV-uptake-report-Climateworks-Centre-August-2022.pdf.

Madanda, R., Musasizi, P. I., Asiimwe, A. T., Matovu, F., Africa, J., & Tickodri-Togboa, S. S. (2013). *Model-based Engineering and Realisation of the KAYOOLA Electric City Bus Powertrain.* Proceedings of the 27th World Electric Vehicle Symposium and Exhibition (EVS27), Barcelona, Spain.

Moeletsi, M. E. (2021). Socio-economic barriers to adoption of electric vehicles in South Africa: Case study of the Gauteng province. *World Electric Vehicle Journal,* 12, 167. https://doi.org/10.3390/wevj12040167.

NLTA (2020). Registration of electric vehicles. https://nlta.govmu.org/Pages/Statistics/Statistics.aspx. Accessed 24 November 2022.

Onai, K., & Ojo, O. (2017). Yehicle-to-grid technology assisted microgrid in Ghana: Opportunities and challenges. *IEEE PES PowerAfrica,* 341–346.

Ramya, K. C., Ramani, J. G., Sridevi, A., Rai, R. S., & Shirley, D. R. A. (2022). Analysis of the different types of electric motors used in Electric Vehicles. In: Kathiresh, M., Kanagachidambaresan, G. R., & Williamson, S. S. (eds) *E-Mobility. EAI/Springer Innovations in Communication and Computing.* Springer. https://doi.org/10.1007/978-3-030-85424-9_3.

Wahab, L., & Jiang, H. (2019). Factors influencing the adoption of electric vehicle: The case of electric motorcycle in northern Ghana. *International Journal for Traffic and Transport Engineering,* 9 (1), 22–37.

Wood, A. E., & Mattson, C. A. (2016). Design for the developing world: Common pitfalls and how to avoid them. *Journal of Mechanical Design,* 138 (30), 031101. https://doi.org/10.1115/1.4032195.

Xu, B., Sharif, A., Shahbaz, M., & Dong, K. (2021). Have electric vehicles effectively addressed CO2 emissions? Analysis of eight leading countries using quantile-on-quantile regression approach. *Sustainable Production and Consumption,* 27, 1205–1214.

13

SHIFTING GEARS TOWARD NET ZERO

Exploring user acceptance of shared automated vehicles

Patrick Dichabeng, Natasha Merat and Gustav Markkula

Introduction

Transportation significantly contributes to global greenhouse gas emissions, impacting climate change and the race towards achieving net zero. As countries worldwide strive to reduce their carbon footprint, finding sustainable and efficient transportation solutions becomes increasingly important. Shared automated vehicles (SAVs) have emerged as a promising alternative to private cars, offering the potential to revolutionise urban mobility, decrease traffic congestion (Dia et al., 2017), and reduce emissions. By enabling a shift from private vehicle ownership to a shared and more efficient transportation system, SAVs can play a critical role in the global pursuit of net zero emissions.

This chapter explores the factors influencing user acceptance of Society of Automotive Engineers (SAE) Level 4 and 5 SAVs, focusing on the attitudes, perceptions, and preferences of private car owners and drivers in the United Kingdom. Understanding these factors is essential to encourage the adoption of SAVs and facilitate the transition toward sustainable transportation systems. Previous acceptance research has found this subset of users to be the least likely to use SAVs (Kyriakidis et al., 2015). By sharing personal research experiences and reflections on the study by Dichabeng et al. (2021), this chapter aims to provide insights into the challenges and opportunities faced in the journey toward net zero and inspire future researchers to contribute to the global effort in tackling climate change.

The context of the research on shared automated vehicles is situated within the broader pursuit of sustainable transportation solutions that can mitigate the environmental impacts of private car usage. With advancements in autonomous vehicle technology, shared automated vehicles can transform

DOI: 10.4324/9781003380566-18

how people travel, presenting a viable alternative to private car ownership (Merat et al., 2017; Nordhoff et al., 2019). The widespread adoption of SAVs can result in more efficient use of resources, reduced traffic congestion, and lower greenhouse gas emissions, contributing to the global effort towards achieving net zero emissions. In this context, the research focused on understanding the factors influencing the acceptance of SAVs, particularly among private car owners and drivers. This large subset of users, who are least likely to use SAVs, is crucial for achieving a significant modal shift towards shared automated mobility. Exploring their attitudes, perceptions, and preferences towards SAVs allows for identifying potential barriers and enablers that can shape the development and deployment of these vehicles to cater to user needs and expectations.

During the research, qualitative data were collected by conducting a focus group study with drivers in the United Kingdom. The research aims to generate insights into the factors driving the acceptance of SAVs among private car owners. These insights can inform policymakers, automotive manufacturers, and service providers in developing SAVs and related infrastructure to help facilitate a smoother transition towards a sustainable transportation system that supports the global pursuit of net zero emissions.

Research background and motivation

Researching Shared Automated Vehicles began with a deep curiosity about how autonomous technology could change transportation systems to address pressing urban mobility challenges and environmental concerns. The aim was to explore innovative and sustainable transportation solutions for cities transitioning towards net zero emissions while enhancing the quality of life for citizens. With privately owned vehicles contributing heavily towards problems such as traffic congestion, air pollution, and inefficient land use in urban environments, SAVs present a promising possibility. They not only offer efficient on-demand mobility services that have the potential to be safer (Piao et al., 2016) but could also promote equitable access for all society members regardless of driving capabilities or disability status (Alessandrini et al., 2015). The investigations into SAVs have been inspired by the many successful pilot projects highlighting their benefits and numerous technological advancements in this field. Nevertheless, it is equally important to remember how human-related factors affect user acceptance when considering the broader adoption of SAVs in our communities now and in the future (Madigan et al., 2017; Nordhoff et al., 2017).

The focus was on uncovering valuable insights related to user acceptance when developing and implementing advanced SAV systems that cater to their preferences. The research considered various areas, including service quality issues, trust factors involved in adapting new technologies such as SAVs, and price values associated with using them responsibly within shared spaces. It ultimately shapes up into developing a valuable conceptual framework for future work exploring the adoption of SAVs among potential users. The commitment was towards making significant strides in the field of SAV systems. The goal was to contribute towards solutions that tackle environmental and urban mobility challenges and pave the way for a more sustainable, inclusive, and efficient transportation future.

Importance of understanding user acceptance and preferences for SAVs

As already stated, Shared Automated Vehicles have the potential to revolutionise urban mobility by providing sustainable and efficient transportation options that are environmentally friendly while reducing traffic congestion and reliance on personally owned vehicles. In achieving net zero goals today, a seamless transition from traditional modes of transport becomes imperative; thus, understanding user acceptance toward this new technology becomes critical for successful adoption in future years. To inform policymakers and manufacturers, awareness of factors influencing user acceptance helps tailor design strategies that better cater for optimal impact outcomes related to fuel efficiency targets set globally. One of the most crucial aspects of promoting SAVs is addressing users' concerns and expectations to achieve high acceptance rates. Previous studies have shown that factors such as trust in technology, service quality, affordability, and perception of the shared space can influence user acceptance significantly.

However, researchers must also focus on specific user groups to understand their unique preferences. Private vehicle owners, who are generally hesitant about SAV adoption, have distinct needs that require careful consideration while developing transportation solutions such as SAVs. They derive a sense of identity, autonomy, or enjoyment from car ownership. Therefore, by identifying these user-specific barriers and social or behavioural challenges in advance, researchers can help reduce obstacles to SAV adoption. It is also essential for stakeholders to work together towards developing strategies that build positive attitudes among users.

Understanding user acceptance helps achieve ambitious net zero goals and fosters sustainable transportation choices. Collaboration between stakeholders and researchers is imperative when seeking to understand elements that influence user acceptance towards self-autonomous vehicles. With this newly acquired knowledge, approaches and adaptations may be made to ensure that delivered products are highly targeted, considering specific user preferences to foster adoption rates. Enabling the adoption and promotion of SAVs holds promise not solely because they lead towards emergent and eco-friendly advancements within transport infrastructure. However, it helps support broader impacts on curbing greenhouse emissions by promoting carbon-neutral solutions for a greener future.

Research method

The study employed an online asynchronous focus group method using focusgroupit.com, enabling flexible participation without real-time interaction (Dichabeng et al., 2021). Twenty-five participants were recruited through various channels and engaged in discussions by answering engagement, exploration, and exit questions (see Figure 13.1). The moderator guided the conversation, ensuring diverse opinions were heard. Data analysis was conducted using NVivo software, applying thematic content analysis with deductive and inductive coding approaches to identify patterns, recurring themes, and insights into user acceptance of SAVs.

Online asynchronous focus group

This study's online asynchronous focus group method offered a novel approach to gathering qualitative data on user acceptance of SAVs. This method allowed participants to engage in discussions and share their opinions without any need to be online simultaneously, providing flexibility and accommodating various schedules. The focus group's main objective was to understand and determine various issues, providing insights into how people perceive the product (SAVs) and the service (mobility) rather than generalising about the broader population.

The online platform used for this study, focusgroupit.com, enabled the creation of open-ended and closed questions as an online forum. Questions were presented along with images and videos, where necessary, to help clarify the topics. Participants were encouraged to answer questions descriptively, engage in discussions with other participants, and build upon each other's responses. To ensure a diverse sample of participants, the recruitment process involved reaching out to potential participants through multiple channels, including the University of Leeds Institute for Transport Studies Driving Simulator database, social media, and a study participant recruitment website. Eligible participants were non-experts in the research

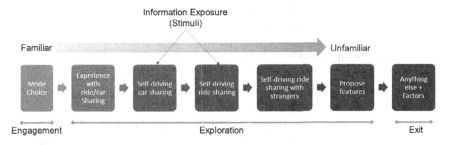

FIGURE 13.1 Flowchart on research method
Source: Dichabeng et al., 2021

topic, owned and used a car every week, had used ridesharing or carsharing services at least once in the past 12 months, and had daily access to a computer or tablet with internet connectivity.

Throughout the study, which ran over five days, participants were presented with engagement and exploration questions designed to introduce the topic, delve deeper into the subject matter, and ensure that all relevant aspects were covered. Exit questions were asked on the last day. The moderator actively guided the discussion, requested clarification on unclear responses, and encouraged participants to reflect on differing opinions. Data analysis was done using NVivo, a qualitative data analysis software, to perform thematic content analysis on the text, identifying patterns, recurring themes, and valuable insights (NVivo, 2018). Both deductive and inductive coding approaches were employed, drawing on existing theories and previous studies while allowing for new themes or categories from the data. Overall, the online asynchronous focus group method provided a valuable means of exploring user acceptance of SAVs, accommodating participants' schedules, and enabling in-depth discussions and exchanging ideas, all contributing to a richer understanding of the factors influencing SAV adoption.

Rationale

The rationale for choosing an online asynchronous focus group method is its ability to foster open and in-depth discussions among participants while offering several advantages over traditional focus groups. Williams et al. (2012) identified the following benefits over conventional face-to-face focus groups:

i Flexibility: Asynchronous focus groups allow participants to engage at their own pace and convenience, providing well-thought-out responses without the pressure of real-time interaction. This flexibility is particularly beneficial for participants with busy schedules or those living in different time zones.

ii Diversity and anonymity: The online platform includes participants from various geographical locations, backgrounds, and experiences, thus ensuring a diverse range of opinions. The anonymity offered by online interactions can also encourage participants to share more honest views, as they might feel less inhibited compared to face-to-face interactions.

iii Reduced social influence: Asynchronous interactions limit the potential for dominant participants to overly influence the group's opinions. This method allows for a more balanced discussion, with all participants having an equal opportunity to contribute to the conversation.

iv Cost and time efficiency: Online asynchronous focus groups eliminate the need for physical venues and travel arrangements, reducing the overall cost and logistical challenges associated with traditional focus groups. This method also allows for more efficient data collection and analysis, as the responses are already digital.

v Rich Data Collection: The combination of open-ended questions and participant interactions results in a wealth of qualitative data that can provide valuable insights into user acceptance of SAVs. This method enables the researcher to better understand participants' attitudes, perceptions, and preferences while exploring emerging themes and patterns.

Novel approaches

The study employed a few novel methodological approaches that set it apart from other research on user acceptance of shared automated vehicles, such as:

i Focus on private car owners: The study targeted private car owners, an underrepresented group in previous SAV acceptance research. By examining this subset of users, we aimed to understand their attitudes and preferences to tailor SAVs to their needs, thus increasing the likelihood of their adoption.

ii Online asynchronous focus group: We utilised an online asynchronous focus group method to allow for more flexible and diverse participation while minimising the influence of dominant participants. This innovative approach enabled us to gather richer and more honest participant data, contributing to a deeper understanding of user acceptance factors.

iii Inductive and deductive coding: During the analysis of responses, we combined deductive coding based on existing factors from the literature with inductive coding, which allowed us to identify new themes and categories emerging from the participants' discussions. This integrated approach comprehensively enabled us to understand the factors influencing SAV user acceptance.

Findings

The study on the acceptance of shared automated vehicles by private car owners and drivers revealed several key findings related to their values, attitudes, and preferences:

i Service quality: Participants emphasised the importance of service quality in their willingness to adopt SAVs. The vehicles' cleanliness, comfort, and convenience were crucial in shaping their perceptions.

ii Trust: Trust emerged as a significant factor influencing acceptance. Participants expressed concerns about the safety and reliability of SAVs and the need to trust the technology and the companies behind it. Additionally, trusting co-passengers and security within the shared space were critical considerations.

iii Price value: The cost of using SAVs was a significant concern for participants. They sought a balance between affordability and the quality of service offered.

iv Shared space quality: The study introduced Shared Space Quality as an essential aspect of SAVs. Participants highlighted the need for well-maintained and comfortable interiors in SAVs, which could influence their decision to use such services.

v Little mention of productive use of travel time: Contrary to previous research, participants in this study rarely mentioned the potential benefits of increased productivity during travel time in SAVs.

These findings provided valuable insights into the values, attitudes, and preferences of private car owners and drivers regarding adopting SAVs. Understanding and addressing these factors could significantly contribute to the successful implementation and widespread acceptance of SAVs in the future.

Challenges and opportunities

Challenges and opportunities existed throughout the research journey on shared automated vehicles acceptance. Some of the most significant challenges when using an online asynchronous focus group included:

i Participant recruitment: Ensuring a diverse sample of private car owners and drivers was challenging. Multiple recruitment strategies, such as using a university database, social media, and a participant recruitment website, were employed to reach potential participants who met the study criteria. However, since online asynchronous focus groups typically use a more significant number of participants than face-to-face groups, a lot more time was spent on screening participants.

ii Maintaining participant engagement: Retaining participant interest and commitment during the online asynchronous focus group over five days was essential. Providing a daily activity guide, offering incentives, and sending reminders through private emails helped maintain participant engagement and encourage active participation. This required daily monitoring of discussions and prompting inactive participants, some of whom did not respond.

iii Analysing qualitative data: Analysing qualitative data from open-ended questions required careful attention and a systematic approach. Using NVivo software for thematic content analysis facilitated the organisation, analysis, and identification of patterns in the data. While a tool such as NVivo is beneficial, analysing this type of qualitative data was time-consuming.

Balancing deductive and inductive coding: Integrating existing theories and previous studies with the newly emerging themes from the data required a balance between deductive and inductive coding. This was achieved by adapting a codebook based on previous research (Nordhoff et al., 2019; Venkatesh et al., 2012) and including additional codes from the data. Despite these challenges, there were also significant opportunities, such as:

i Gaining in-depth understanding: The online asynchronous focus group method allowed for a rich exploration of participants' attitudes, perceptions, and preferences, providing valuable insights into factors influencing SAV acceptance.
ii Novel methodological approaches: The introduction of shared space quality as a component of service quality and the emphasis on trust (including trusting co-passengers and security) contributed to a more comprehensive understanding of user acceptance factors.
iii Developing a conceptual acceptance model: The study's findings contributed to developing a conceptual SAV technology acceptance model, which extended the Unified Theory of Acceptance and Use of Technology (UTAUT2) by Venkatesh et al. (2012).

Overcoming obstacles and embracing opportunities throughout the research journey helped to deepen the understanding of private car owners' and drivers' values, attitudes, and preferences regarding SAVs. This, in turn, contributed to developing effective strategies for promoting SAV adoption and achieving the broader goal of net-zero emissions.

Impact on net zero and personal reflections

This study contributes to the net zero goal by providing valuable insights into the factors influencing private car owners' and drivers' acceptance of shared automated vehicles. Encouraging a shift from private car use to SAVs is essential for reducing greenhouse gas emissions, alleviating traffic congestion, and promoting sustainable urban mobility. By identifying the key factors that impact user acceptance, such as service quality, trust, and price value, the research offers practical recommendations for the design and implementation of SAV services. These recommendations can be used by policymakers, urban planners, and SAV service providers to tailor SAV offerings to meet private car owners' and drivers' needs and preferences, ultimately increasing the likelihood of SAV adoption.

In particular, the research highlights the importance of addressing potential SAV users' mobility and affective needs. By creating a shared space that is comfortable, secure, and trustworthy, and by offering a service that is reliable, efficient, and affordable, future SAV service providers can become a more attractive alternative to private car use. Additionally, developing a conceptual SAV technology acceptance model, which expands on the Unified Theory of Acceptance and Use of Technology (UTAUT2), provides a valuable framework for future research on SAV acceptance. This model can help guide further investigations into user intentions and behaviour, enabling a better understanding of the most effective strategies for promoting SAV adoption. Overall, the research contributes to the net zero goal by providing a foundation for understanding and addressing the barriers to SAV adoption among private car owners and drivers. By encouraging a shift from private cars to SAVs, we can reduce carbon emissions, improve urban sustainability, and move closer to achieving a net zero future.

Milestones achieved throughout the research on SAVs

Throughout the research journey on Shared Automated Vehicles (SAVs), we have experienced various challenges, successes, and personal growth. From the initial stages of identifying the research gap and formulating the research question to designing and conducting the study, analysing the data, and presenting the findings, this journey has been a rewarding and enlightening experience. One of the critical milestones in the study was the successful implementation of the online asynchronous focus group study. This novel methodological approach allowed us to gather in-depth insights from diverse participants, providing valuable data on their attitudes, perceptions, and preferences for SAVs. This method also enabled participants to engage in the study at their convenience, which resulted in richer and more thoughtful responses. Another significant milestone was developing the conceptual SAV technology acceptance model, which expands on the Unified Theory of Acceptance and Use of Technology (UTAUT2). This model is a valuable contribution to the field, offering a framework for future research on SAV acceptance and user behaviour.

Throughout the research journey, we have also gained a deeper understanding of the complexities and nuances of SAV adoption, particularly among private car owners and drivers. We have learned the importance of considering potential users' mobility and affective needs and have developed a greater appreciation for the multifaceted nature of technology acceptance and adoption. The research journey has been an incredible opportunity for growth and development. We have honed our research design, data analysis, and academic writing skills while cultivating perseverance in facing challenges and setbacks. The lessons learned and experiences gained during this journey will undoubtedly serve us well as we continue to explore the fascinating world of SAVs and their potential to contribute to a sustainable and net zero future.

Thoughts on the future of SAVs and their potential role in achieving net zero

The future of shared automated vehicles (SAVs) appears promising and challenging. The ability to improve the transportation sector by reducing greenhouse gases while encouraging a more sustainable environment for tomorrow makes them an intriguing development worth considering. A shift from private ownership towards shared mobility services leveraging SAVs' efficiency could significantly reduce the number of vehicles on roads, with associated emissions declining accordingly. Technologically advanced high-speed SAVs promise better fuel efficiency and promote alternative energy sources, such as electricity, that will help further reduce greenhouse gas concentrations in the atmosphere.

Integrating these revolutionary new concepts within broader transport networks, including public transportation systems, creates a truly smart city landscape promoting sustainable urban environments available for all users while ensuring safety, trust, privacy, and shared space quality, as well as accommodating diversified user preferences as essential factors towards achieving this goal. For self-driving cars such as SAVs to become more common on roads across the globe, key players such as industry leaders, policymakers, and researchers must collaborate on targeted projects to overcome both technological and human obstacles. Embracing electric-powered autonomous vehicles such as SAVs has the potential for a significant impact to reduce harmful gas emissions drastically. This will help provide environmentally friendly alternatives to conventional cars. They are essential in helping humanity meet global goals regarding reducing carbon footprint levels.

Conclusion

This chapter focused on sharing the experience of trying to understand user acceptance of shared automated vehicles using an online asynchronous focus group method. This method allowed for a comprehensive exploration of participants' attitudes, perceptions, and preferences while providing flexibility and convenience to both researchers and participants. The research employed a combination of engagement, exploration, and exit questions to guide the discussions, with the help of focusgroupit.com as the instrument. Data analysis was conducted using NVivo, applying thematic content analysis to identify recurring themes and patterns in participants' responses. Deductive coding was employed, using a codebook derived from existing theories and previous studies, with inductive coding applied when necessary to capture novel themes. This methodological approach enabled the identification of critical factors influencing SAV acceptance among private car owners and drivers, such as service quality, trust, and price value. The insights obtained from the study contributed to the development of a conceptual SAV technology acceptance model, extending the Unified Theory of Acceptance and Use of Technology (UTAUT2).

The research on Shared Automated Vehicles (SAVs) is highly relevant to the broader net zero context, as SAVs have the potential to significantly reduce greenhouse gas emissions, traffic congestion, and reliance on privately owned vehicles. By understanding the factors influencing user acceptance, particularly among private car owners and drivers, the study (Dichabeng et al., 2021) provides insights to inform the design and implementation of SAV services that effectively cater to user needs and preferences. As private car owners are generally less likely to adopt SAVs, the study's focus on this subset of users is crucial in promoting a shift towards more sustainable modes of transportation. By tailoring SAVs to meet mobility and affective needs, policymakers and industry stakeholders can encourage wider adoption of SAVs, thereby contributing to the net zero goal by reducing carbon emissions, optimising transportation efficiency, and fostering more sustainable urban environments.

Throughout the research journey on SAVs, we have delved deep into understanding user acceptance, preferences, and the factors that can drive the adoption of this innovative mode of transportation. The process has been filled with challenges and opportunities, allowing us to learn, adapt, and grow as researchers. We have employed novel methodological approaches, which provided unique insights into private car owners' and drivers' attitudes and preferences. The potential impact of this work on achieving net zero is significant, as it sheds light on the factors that can encourage a shift from private cars to SAVs. The study contributes valuable knowledge that can be used by policymakers, city planners, and industry stakeholders to design and implement SAV services that appeal to a broader audience, especially the private car users who are least likely to adopt such services. By helping to facilitate this transition, this study plays a critical role in reducing greenhouse gas emissions, alleviating traffic congestion, and promoting sustainable urban living. The research journey has contributed to the broader understanding of SAVs and their role in achieving net zero. It has also served as a testament to the power of determination and curiosity in driving meaningful change. As we explore the future of SAVs and their potential impact on urban mobility, our work will undoubtedly inspire others to strive towards creating a more sustainable world.

References

Alessandrini, A., Campagna, A., Site, P. D., Filippi, F., & Persia, L. (2015). Automated vehicles and the rethinking of mobility and cities. *Transportation Research Procedia*, 5, 145–160. https://doi.org/10.1016/j.trpro.2015.01.002.

Dia, H., & Javanshour, F. (2017). Autonomous shared mobility-on-demand: Melbourne pilot simulation study. *Transportation Research Procedia*, 22, 285–296. https://doi.org/10.1016/j.trpro.2017.03.035.

Dichabeng, P., Merat, N., & Markkula, G. (2021). Factors that influence the acceptance of future shared automated vehicles – a focus group study with United Kingdom drivers. *Transportation Research Part F: Traffic Psychology and Behaviour*, 82, 121–140. https://doi.org/10.1016/j.trf.2021.08.009.

Kyriakidis, M., Happee, R., & de Winter, J. C. F. (2015). Public opinion on automated driving: Results of an international questionnaire among 5000 respondents.

Transportation Research Part F: Traffic Psychology and Behaviour, 32, 127–140. https://doi.org/10.1016/j.trf.2015.04.014.

Madigan, R., Louw, T., Wilbrink, M., Schieben, A., & Merat, N. (2017). What influences the decision to use automated public transport? Using UTAUT to understand public acceptance of automated road transport systems. *Transportation Research Part F: Traffic Psychology and Behaviour*, 50, 55–64. https://doi.org/10.1016/j.trf. 2017.07.007.

Merat, N., Madigan, R., & Nordhoff, S. (2017). Human factors, user requirements, and user acceptance of ride-sharing in automated vehicles. In *ITF Discussion Papers: International Transport Forum*.

Nordhoff, S., van Arem, B., Merat, N., Madigan, R., Ruhrort, L., Knie, A., & Happee, R. (2017, 19–22 June). *User Acceptance of Driverless Shuttles Running in an Open and Mixed Traffic Environment*. Paper presented at the 12th ITS European Congress, Strasbourg, France.

Nordhoff, S., de Winter, J., Payre, W., van Arem, B., & Happee, R. (2019). What impressions do users have after a ride in an automated shuttle? An interview study. *Transportation Research Part F: Traffic Psychology and Behaviour*, 63, 252–269. https://doi.org/10.1016/j.trf.2019.04.009.

NVivo. (2018). Introducing NVivo. https://www.qsrinternational.com/nvivo/what-is-nvivo.

Piao, J., McDonald, M., Hounsell, N., Graindorge, M., Graindorge, T., & Malhene, N. (2016). Public views towards implementation of automated vehicles in urban areas. *Transportation Research Procedia*, 14, 2168–2177. https://doi.org/10.1016/j.trpro.2016.05.232.

Venkatesh, V., Thong, J. Y. L., & Xu, X. (2012). Consumer acceptance and use of information technology: Extending the Unified Theory of Acceptance and Use of Technology. *MIS Quarterly*, 36 (1), 157–178. https://doi.org/10.2307/41410412.

Williams, S., Clausen, M. G., Robertson, A., Peacock, S., & McPherson, K. (2012). Methodological reflections on the use of asynchronous online focus groups in health research. *International Journal of Qualitative Methods*, 11 (4), 368–383. https://doi.org/10.1177/160940691201100405.

PART V

FASHION AND SERVICE SECTORS

14

INTEGRATING SUSTAINABLE PRODUCT-SERVICE SYSTEM AND SERVICE DESIGN FOR NET ZERO IN DEVELOPING ECONOMIES

Yaone Rapitsenyane

Design for environment at childhood

As a typical African boy, growing up was characterised by homemade toys. The playground was filled with clay-made cows and cars, mud huts and footballs made of recycled polythene bags. Our toy cars were also made of soft steel wire cut-offs, with wheels made of recycled soda cans. We built kites from old polythene bags and sticks and controlled them so that they wouldn't fly away from us using strings. We would reuse packaging cans of tuna and a high-tension wire or rope for our fixed telephone lines. It was a wonderful world of quick solutions to building exciting products, which brought us joy in the playground. It was never evident that I would become a product designer. I started doing design and technology at junior school and senior secondary school, after which I studied a design and technology undergraduate programme at the University of Botswana.

A trajectory of design for sustainability

As an undergraduate in design and technology, I became fascinated by problem-solving and was particularly interested in one module called Design for Sustainable Rural Development. I was intrigued by the works of Victor Papanek and his concept of how design relates to people (Papanek & Fuller, 1972) and E. F. Schumacher's concept of making technology, which is synergistic with the human beings it is meant for (Schumacher & Beautiful, 1977). Rachael Carson's Silent Spring (Carson, 1962) was an eye-opener on the effects of environmental pollution and adverse effects of human activity on the environment, especially the use of fertilisers and pesticides on water and the life of species around agricultural lands. The central theme of these

DOI: 10.4324/9781003380566-20

perspectives became a whole as I read the Brundtland Commission's report (WCED, 1987) and their definition of sustainable development as development that meets the needs of the present generation without compromising the ability of the future generation to meet their needs. The role of design in sustainable development is articulated in terms of the redesign of products for better resource usage regarding energy and materials.

This sustainability knowledge influenced my graduation project. I focused on how small-scale farmers in a typical developing country such as Botswana can care for their citrus fruit trees in high-temperature regions. Using design research methods, I conducted user research to understand the needs of small-scale citrus farmers regarding the problems they encounter during hot seasons and the technologies they use to solve them. Primitive solutions emerged as a standard in my research findings. During the interviews with the farmers, this was attributed to the high costs of proper technological equipment to provide a greenhouse environment for the trees to protect them from heat stress. My proposal was an intermediate technology triggered by probes, which, as the topsoil around the trees becomes dry, intermittently produce mist, reducing the high evaporation rates which lead to dryness. The product was made of recyclable polypropylene with a semi-knockdown modular design to aid in packing it away when it is not needed. Even though the objective of the graduation project was a culmination of all skills for design and prototyping in the undergraduate programme, the product was heavily influenced by the socio-economic and environmental considerations of the Design for Sustainable Rural Development module. This was the beginning of my design for sustainable growth and career influence.

Design for environment in early career life

My first job was as a design and technology teacher, which I did for only 16 months before I was admitted into an MSc in Sustainable Product Design at Loughborough University. This further training path landed me in an academic career where I developed sustainable design as my speciality and a broad area of research, later shaping my research engagement after my doctoral studies in the areas of product-service system design for sustainability and service design. As a young academic teaching and learning how to do research in Design for the Environment, I invested a lot in learning and teaching environmental sustainability through project-based learning. Design for environment or eco-design is the systematic identification and integration of environmental considerations into the New Product Development process, considering the entire life cycle of a product at the fuzzy front end of the design process (Deutz et al., 2013; Park & Tahara, 2008; Bovea & Pérez-Belis, 2012).

As part of teaching sustainability courses, the eco-design practice was inculcated in students practically using eco-design tools. Design activities were modelled around how to improve the environmental performance of products and how to inform the redesign of basic consumer products from the perspective of eco-design. These activities evolved through a structured approach as the students also learn the design process (Deutz et al., 2013; Vallet et al., 2013), as these eco-design activities are usually synchronised with the objectives of the New Product Development process (Vallet et al., 2014). In the process of teaching design for the environment, I was conducting research to look at sustainability issues in the design curriculum of institutions in southern Africa (Moalosi et al., 2010) and how design has embraced eco-design culminating in the sustainable design (Rapitsenyane, 2011).

Product-service system research

My PhD research was on product-service systems as a sustainable business strategy for non-design-led small and medium-sized enterprises (SMEs) in the manufacturing industry. The fascinating fact about the product-service system is the implications for manufacturing companies in the face of transitions in the global economy. Transitions in economic activities from single-product transactions to co-creating experiences (De Lille et al., 2012) will, over time, render traditional manufacturing obsolete as far as meeting present user needs. Developments in creating a sustainable manufacturing environment vary from waste reduction to complex closed-loop systems approaches (Jayal et al., 2010) and often require companies to possess additional capabilities to make necessary adjustments. These adjustments often require changes in business strategies to enable the companies to adapt to new business environments and differentiate their offerings. A culture of constant innovation can enable this adaptation, especially where the goal is to decouple economic success from resource consumption and achieve service-oriented differentiation through strategies encouraging a cyclic flow of resources and extended product life cycles (Rapitsenyane & Bhamra, 2013).

There are eight archetypical models of S.PSS identified by Tukker (2004). The same classification can be found in Hernandez-Pardo (2012) and Rapitsenyane (2014). These are classified under three main categories: product-oriented (the user retains ownership, and the provider offers after-sales support services), use-oriented, and results-oriented (where the provider retains ownership of the product, and the user pays for services they get from the product). Product-oriented S.PSS include product-related services, advice, and consultancy. Use-oriented S.PSS include product lease, product renting or sharing, and product pooling. Results-oriented models include activity management/outsourcing, pay-per-service, and functional units.

S.PSS has been widely researched in terms of tools and methods, classifications, applications, benefits, drivers, and barriers, including best practices (Ang et al., 2012; Baines et al., 2007; Mont, 2002; Kang & Wimmer, 2009; Hernandez-Pardo et al., 2013; Tukker, 2004). More recently, product-service system business models have been explored to enhance circularity in supply chains through how value is created in PSS business models (Yang et al., 2018; da Costa Fernandes et al., 2020). These transitions need to be supported in terms of building capacity for providers from traditional manufacturing. While a considerable amount of work has been done with multinational corporations to offer methodological support (Gomes et al., n.d.; Kang & Wimmer, 2009); criticised for its tool's intensity (van Halen et al., 2005), an equally considerable amount of work has not been done for SMEs. Digitisation of the transition to a product-service system to aid companies in being better able to compete (Solem et al., 2022) adds another complexity for SMEs.

SMEs need more investments to support professional design practice and advanced digital technologies (Moultrie et al., 2006). However, SMEs comprise the more considerable proportions of companies in many contexts. For example, in Italy, SMEs make up 95% of all companies in the furniture industry (Dell'Era et al., 2018). Most SMEs access design capabilities through collaborations with designers to contribute to their product portfolios. An interesting study by Hernandez-Pardo (2012) proposed a framework to guide the design of sustainable PSS in SMEs in Colombia, with design and Information Communication Technology as an integrated context of sustainable business development. Another design approach is proposed by De Lille et al. (2012) for traditional product-oriented manufacturers to shift towards an S.PSS approach, but the relevant capabilities to be developed, especially for non-design-led companies, still need to be identified.

In product-service systems literature synonymous with servitisation, the focus has been on developing tools and methods (Morelli, 2006; Hernandez-Pardo, 2012; Finken et al., 2013). However, some of these studies provide awareness of capabilities relevant to companies for solving user needs in an S.PSS approach (Finken et al., 2013). How companies, especially SMEs, can consciously develop capabilities to operate as S.PSS providers have yet to be studied. The complexity involved in engaging non-design-led SMEs in this process makes S. PSS a complex paradox yet an advantage if both design and S.PSS capabilities were to be developed simultaneously to aid SMEs towards competitiveness.

The literature on S.PSS acknowledges and mentions that the concept can give companies a competitive edge and includes it under the benefits of S. PSS (Hernandez-Pardo et al., 2012; Martinez et al., 2010). Indicators of this competitive advantage seem not to be coming out clear, though this has not been a focus in any of the studies. An attempt, however, is made by Tukker (2004) and Meier et al. (2010). Tukker assesses the extent to which a S.PSS can be a worthwhile investment and analyses critical economic elements of each of the eight S.PSS archetypical models he identified. The analysis was based on four elements identified: the market value of the S.PSS; production costs of the S.PSS; investment needs/capital needs for S.PSS production; and the ability to capture the value present in the value chain now and in the future (Tukker, 2004).

Tukker's approach to assessing competitiveness is price-related as opposed to looking at non-price factors impacting the competitiveness of companies. Meier et al. (2010) refer to Roy and Chevuru (2009) for their framework on competitive S.PSS and connect competitiveness and drivers towards S.PSS in Roy and Chevuru's framework. Based on the drivers (customer affordability, revenue generation opportunity, global competition, technology development, environmental sustainability) acting on the commercial environment, the framework is critical to customer value and business success. The framework's inclination towards big companies renders it less relevant for SMEs.

From my PhD research, the service component of S.PSS is an opportunity for SMEs to improve their competitiveness, as differentiation through services becomes more critical than through product offerings. A mindset change is crucial as customers need to be convinced of the value of satisfaction through non-product ownership, which was previously achieved through owning a product. Understanding users' values and behaviours is critical to enable manufacturers to channel their S.PSS ideas towards acceptable offerings in a service context. This may not be a concern for countries in Europe, such as Norway, where a service culture is already at its maturity stage (Gloppen, 2009) or in developed countries, as generally argued by Baines et al. (2007). From the manufacturers' point of view, this is a cause for concern in contexts where a service culture is still developing. A feeling of scepticism about acceptance of their S.PSS offerings is a barrier to adoption.

The thesis proposed strategies to address the mindset issue of manufacturers being service providers. Responsiveness of these strategies to contextual dynamics is crucial for S.PSS to be effectively adopted, lest it faces the risk of rejection and being viewed as an irrelevant and foreign concept. In my thesis, design can play a pivotal role in S.PSS adoption when used as a strategy to address the competitiveness concerns of SMEs. The traditional interpretation of design creativity and imagination makes products look nice and different. This is mainly from the ability of design to look for possibilities that may not have been explored before. Transferring this capability in the business context means focusing on the product and the entire innovation process, especially in the early stages where identifying business opportunities as an entrepreneur is critical. Scanning the external environment requires some creative intelligence. Even though business tools are still needed for this process, a design attitude is needed to provide a unique mindset to problem-solving. The idea behind business innovation is to have customers who can buy and use the offering. This attitude aims to make the world a better place, and using it in manufacturing SMEs to identify S.PSS opportunities is critical to instil the ability to think differently. The engagement of SMEs with designers should be an informal educative process for SMEs to learn S.PSS development processes and their value to the company and stakeholders. This should offer better alternatives to existing company approaches that view the external environment as buyers only rather than as co-creators of value. The significant impact to be made through this engagement will depend on the designers' ability to build confidence in SMEs about their involvement with their businesses.

The thesis explored a contextually appropriate approach for manufacturing SMEs in Botswana to address their competitiveness needs by shifting to a sustainable product-service system supported by developing design capabilities. The engagement of SME participants with designers in interactive workshops has demonstrated and identified factors that can contribute to this shift. These factors have contributed to the Design Capabilities for Product-Service System (DeCap PSS) process shown in Figure 14.1.

Factors contributing as stages of the DeCap S.PSS process are: (i) identifying value to initiate engagement, (ii) building understanding, (iii) reflecting on familiar experiences, (iv) empowering and coordinating, (v) defining outlook of the organisation, and (vi) proposing added advantage. The first three stages are primarily concerned with developing design creativity, and the later design strategy over a trajectory that involves design as an engaging activity that connects creativity to strategy. Similarities can be drawn between the six stages of the DeCap S.PSS process and the stages of changing an organisation into a S.PSS provider proposed by De Lille et al. (2012). In both cases, the emphasis is on a balance between understanding business thinking and understanding the organisation so that there is a way of manoeuvring S.PSS ideas towards implementation and design capabilities to support innovation and changes from product to service orientation.

Reporting back to work from my doctoral studies, I joined an ongoing research project in the Learning Network for Sustainable Energy Systems (LeNSes) on sustainable product-service systems and renewable and distributed energy systems. LeNSes was a collaborative project between universities in Europe and Africa, co-funded by Erasmus+ and facilitated by the

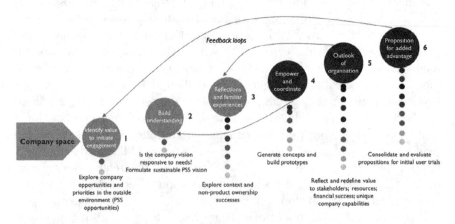

FIGURE 14.1 Design capabilities for product-service systems

Learning Network on Sustainability International (2022). The LeNSes project aimed to promote a new generation of design educators and researchers capable of designing sustainable energy systems for all in Botswana, Kenya, South Africa, and Uganda, explicitly focusing on Sustainable Development Goal 7: Affordable and Clean Energy. LeNSes produced an interesting web-package platform providing learning resources supported by a Creative Commons licence free of charge for users, teachers, and learners. From experiences in the LeNSes project, sustainable product-service systems and renewable energy systems were integrated into the University of Botswana curriculum, topics that I still teach to this day.

Service design research

During my doctorate studies, I participated in my first-ever global service jam at Loughborough University in 2012. At the jam, I had an experience with the service design process and proposing a service within 48 hours in a multidisciplinary design team. After my doctoral studies, I pioneered the global service jam in Botswana, with the first jam in 2015. The objective was to promote the field of service design in Botswana and demonstrate how services could be better designed and delivered with a design influence. Service design aims to improve the customer experience and the interactions between the provider and the customer by designing and marketing benign end-to-end services (Blomkvist et al., 2010; Stickdorn et al., 2018).

As part of service design research in 2018 through the Enhancing Engineering Education project funded by the Royal Academy of Engineering (RAE), I did a secondment in a high-end restaurant to explore how they could use service design to enhance customer experience and increase business value. The primary purpose of the work done with the RAE funding was to enhance academics' participation in shaping the Botswana industry through secondment placements. My research aimed to identify service design strategies for SMEs in the high-end restaurant business.

The SME was based in a town in the northern part of Botswana. The SME wanted to turn around the restaurant side of the business and had provided a brief to the RAE project coordinators as 'rethinking the business concept'. One of the valuable additional user-centric aspects of service design is using tools through a sequential process that uncovers pain points and wow moments in a service (Zomerdijk & Voss, 2010; Rejikumar et al., 2022). As a design educator, I was seconded to the company for two months, split into two sessions of four weeks each. The purpose of the placement was to immerse the designer in the restaurant's business activities and to share design experiences from the academic's perspective within a business environment. It is imperative to involve users in the co-development of services to enhance the analysis of user experiences and critical incidents (Trischler & Scott, 2016). The SME owner/manager sought sustainable design proposals for quick business wins. These were possible if the focus was devoted to front-end improvements, though they are usually unsustainable (Kuk & Janssen, 2013). From the experiences of the secondment, these were to be implemented through a product/service business strategy. The data were collected through observations of customer visits, recorded in terms of what they order, when they order, sit in, or take out, and waiting time and payment method. The data was also collected by observing front-stage and backstage staff actions and interactions with customers and backstage processes, reviewing documents, immersion in the service as backstage and front-stage staff, and interviews with the owner/manager and two staff members. The data were analysed qualitatively through thematic analysis to identify opportunities for rethinking the business concept captured by the owner's brief.

The resultant proposals were front-end heavy but passively impacted the customer experiences, streamlining the services provided to only those in demand. Service advertisements immediately raised positive awareness about the service through physical representation and demonstration of the service encounter (Mortimer, 2008). This strategy was effective through service advertising campaigns at strategic spots in the town. Coming from a developing country, my current service design research in the medical industry explores how medical service ecosystems and high technology can be used for local services traditionally provided internationally by sending patients to developed countries at very high medical costs.

Sustainability and the technology-driven future

The fourth industrial revolution is providing technology that takes over tasks traditionally done by human beings. Human activity has adversely affected our planet and all its inhabitants, grossly affecting the quality of life for future generations. Are we going into more destructive pathways? Products have become smaller and smaller as technology improved and dematerialisation became achievable. However, miniaturised technology products are still dependent on rare mineral resources, the shortage of which in some cases still halts manufacturing such products as vehicles. Electric vehicles are one of the Industry 4.0 technologies. The principle behind electric vehicles is that they will not emit carbon dioxide (CO_2), and thus positively contribute to the overall reduction of CO_2 emissions and avoidance of temperature rise. On the flip side, many developing countries still depend on burning coal to produce their energy, especially electricity. Using electric vehicles powered by fossil fuels will still increase CO_2 emissions. Distributed manufacturing has also led to finished goods being transported over long distances on airfreight or sea freight. These modes of transport still use fossil-based fuels.

Lately, there have been studies conducted on sustainable fuel options. These fuel options still contribute about 30% to the total fuel used in transportation today. The World Wide Web has revolutionised how products and services are co-produced and co-consumed. Virtual co-production and co-consumption platforms are on the rise to allow collaborations across cultures, industries, and businesses in various geographic locations. Sustainable product-service systems have also transcended regional boundaries, making them unsustainable. Sustainability should be explored in the context of high technology for high impact on a global scale. The biggest concerns in sustainability, energy, and materials should drive the next big agenda with a sustainability goal in mind.

References

Ang, G. C. J., Baines, T., & Lightfoot, H. (2012, October). *A methodology for adopting product-service systems as a competitive strategy for manufacturers.* In Proceedings of the 2nd CIRP IPS2 Conference 2010, 14–15 April, Linköping, Sweden (No. 077, pp. 489–496). Linköping University Electronic Press.

Baines, T. S., Lightfoot, H. W., Evans, S., Neely, A., Greenough, R., Peppard, J., ... & Wilson, H. (2007). State-of-the-art in product-service systems. *Proceedings of the Institution of Mechanical Engineers, Part B: Journal of Engineering Manufacture*, 221(10), 1543–1552.

Blomkvist, J., Holmlid, S., & Segelström, F. (2010). Service design research: Yesterday, today, and tomorrow. In *This Is Service Design Thinking: Basics – Tools – Cases* (1st ed., pp. 308–315). Retrieved from http://urn.kb.se/resolve?urn=urn:nbn: se:liu:diva-80953.

Bovea, M. D., & Pérez-Belis, V. (2012). A taxonomy of ecodesign tools for integrating environmental requirements into the product design process. *Journal of Cleaner Production*, 20(1), 61–71.

Carson, R. (1962). *Silent Spring*. Houghton, Mifflin, Harcourt.

da Costa Fernandes, S., Pigosso, D. C., McAloone, T. C., & Rozenfeld, H. (2020). Towards product-service system oriented to a circular economy: A systematic review of value proposition design approaches. *Journal of Cleaner Production*, 257, 120507.

De Lille, C. S. H., Roscam Abbing, E., & Kleinsmann, M. S. (2012). *A designerly approach to enable organizations to deliver product-service systems*. In International DMI Education Conference: Design Thinking: Challenges for Designers, managers and Organizations, 14–15 April 2008, Cergy-Pointoise, France. DMI.

Dell'Era, C., Magistretti, S., & Verganti, R. (2018). Exploring collaborative practices between SMEs and designers in the Italian furniture industry. In *Researching Open Innovation in SMEs* (pp. 307–345).

Deutz, P., McGuire, M., & Neighbour, G. (2013). Eco-design practice in the context of a structured design process: an interdisciplinary empirical study of UK manufacturers. *Journal of Cleaner Production*, 39, 117–128.

Finken, K. H., McAloone, T. C., Avlonitis, V., Garcia I Mateu, A., Andersen, J. A. B., Mougaard, K., Neugebauer, L. M., & Hsuan, J. (2013). *PSS Tool Book: A Workbook in the PROTEUS Series*. Technical University of Denmark (DTU).

Gloppen, J. (2009). Perspectives on design leadership and design thinking and how they relate to European service industries. *Design Management Journal*, 4(1), 33–47.

Gomes, N. S., Vasques, R. A., Chang, A., Kramer, A., Nájera, A. H., & Soto, S. L. (n.d.). An exploratory study on the design of a product service system: The case of Soliforte. [unpublished paper] Retrieved from: https://www.academia.edu/download/31675436/Gomes_Vasques_et_al_An_exploratory_study_on_the_design_of_a_PSS_the_case_of_Soliforte.pdf.

Hernandez-Pardo, R. (2012). *Designing Sustainable Product Service Systems: A Business Framework for SME Implementation*. Doctoral dissertation, Loughborough University.

Hernandez-Pardo, R. J., Bhamra, T., & Bhamra, R. (2013). Designing sustainable product service systems in SMEs. *The International Journal of Design Management and Professional Practice*, 6(4), 57.

Jayal, A. D., Badurdeen, F., Dillon Jr, O. W., & Jawahir, I. S. (2010). Sustainable manufacturing: Modeling and optimisation challenges at the product, process and system levels. *CIRP Journal of Manufacturing Science and Technology*, 2(3), 144–152.

Kang, M., & Wimmer, R. (2009). Product Service Systems Beyond Sustainable Products Case Study of Prefabricated Unit House Reuse System. *Proceeding of Sustainable Innovation 03-Creating Sustainable Products, Services and Product-Service-Systems*, 18–22.

Kuk, G., & Janssen, M. (2013). Assembling infrastructures and business models for service design and innovation. *Information Systems Journal*, 23(5), 445–469.

Martinez, V., Bastl, M., Kingston, J., & Evans, S. (2010). Challenges in transforming manufacturing organisations into product-service providers. *Journal of Manufacturing Technology Management*, 21(4), 449–469.

Meier, H., Roy, R., & Seliger, G. (2010). Industrial product-service systems – IPS2. *CIRP Annals*, 59(2), 607–627.

Moalosi, R., Rapitsenyane, Y., & M'Rithaa, M. (2010). *An analysis of sustainability issues in southern African design institutions' programmes*. Proceedings of Sustainability in Design: Now! LeNS Conference, 29 September–1 October, Bangalore, India.

Mont, O. K. (2002). Clarifying the concept of product-service system. *Journal of Cleaner Production*, 10(3), 237–245.

Morelli, N. (2006). Developing new product service systems (PSS): methodologies and operational tools. *Journal of Cleaner Production*, 14(17), 1495–1501.

Mortimer, K. (2008). Identifying the components of effective service advertisements. *Journal of Services Marketing*, 22(2), 104–113.

Moultrie, J., Clarkson, P. J., & Probert, D. (2006). A tool to evaluate design performance in SMEs. *International Journal of Productivity and Performance Management*, 55(3/4), 184–216.

Papanek, V., & Fuller, R. B. (1972). *Design for the Real World*. Pantheon.

Park, P. J., & Tahara, K. (2008). Quantifying producer and consumer-based eco-efficiencies for the identification of key eco-design issues. *Journal of Cleaner Production*, 16(1), 95–104.

Rapitsenyane, Y. (2011). *From eco-design to sustainable design: Experiences from the University of Botswana*. In Proceedings of the Nairobi International Design Conference, 25–27 May, National Museum of Kenya, Nairobi.

Rapitsenyane, Y. (2014). *Supporting SMEs Adoption of Sustainable Product Service Systems: A Holistic Design-led Framework for Creating Competitive Advantage*. Doctoral dissertation, Loughborough University.

Rapitsenyane, Y., & Bhamra, T. (2013). The place of sustainability through Product Service Systems in manufacturing SMEs in Botswana: A Delphi study. In Proceedings of the 16th Conference of the European Roundtable on Sustainable

Consumption and Production (ERSCP) & 7th Conference of the Environmental Management for Sustainable Universities (EMSU), 4–7 June, Istanbul.

Rejikumar, G., Aswathy, A. A., Jose, A., & Sonia, M. (2022). A collaborative application of design thinking and Taguchi approach in restaurant service design for food wellbeing. *Journal of Service Theory and Practice*, 32(2), 199–231.

Roy, R., & Cheruvu, K. S. (2009). A competitive framework for industrial product-service systems. *International Journal of Internet Manufacturing and Services*, 2(1), 4–29.

Schumacher, E. F., & Beautiful, S. I. (1977). *Economics as if People Mattered*. Thesis.

Solem, B. A. A., Kohtamäki, M., Parida, V., & Brekke, T. (2022). Untangling service design routines for digital servitization: Empirical insights of smart PSS in maritime industry. *Journal of Manufacturing Technology Management*, 33(4), 717–740.

Stickdorn, M., Hormess, M. E., Lawrence, A., & Schneider, J. (2018). *This Is Service Design Doing: Applying Service Design Thinking in the Real World*. O'Reilly Media, Inc.

Trischler, J., & Scott, D. R. (2016). Designing public services: The usefulness of three service design methods for identifying user experiences. *Public Management Review*, 18(5), 718–739.

Tukker, A. (2004). Eight types of product-service systems: Eight ways to sustainability? Experiences from SusProNet. *Business Strategy and the Environment*, 13(4), 246–260.

Vallet, F., Eynard, B., & Millet, D. (2014). Proposal of an eco-design framework based on a design education perspective. *Procedia CIRP*, 15, 349–354.

Vallet, F., Eynard, B., Millet, D., Mahut, S. G., Tyl, B., & Bertolucci, G. (2013). Using eco-design tools: An overview of experts' practices. *Design Studies*, 34(3), 345–377.

WCED. (1987). World commission on environment and development. *Our Common Future*, 17(1), 1–91.

Yang, M., Smart, P., Kumar, M., Jolly, M., & Evans, S. (2018). Product-service systems business models for circular supply chains. *Production Planning & Control*, 29(6), 498–508.

Zomerdijk, L. G., & Voss, C. A. (2010). Service design for experience-centric services. *Journal of Service Research*, 13(1), 67–82.

15

A SUSTAINABLE FASHION SAFARI

Sophia N. Njeru

Professional education in design

I am a Kenyan, and pursued my first two degrees at Kenyatta University: a Bachelor of Education (Home Economics) and a Master of Science (Textile Science and Design). I earned a PhD (Design) at Maseno University. Both universities are in Kenya. My area of specialisation is fashion design. My PhD thesis supervisors were Prof Susan Abong'o and Prof Caleb Okumu. The latter passed on two months before my *viva voce*. I grieved and wondered how I would perform at the *viva voce*. However, all went well, and I was the first to graduate with a PhD in Design at the University.

Sustainable fashion research

Safari is a Kiswahili word for journey, expedition, or adventure. My sustainable fashion *safari* is described chronologically. The *safari* started in 2000, albeit without knowing it. I would buy second-hand curtains in open markets and construct unique scatter cushion covers for clients as a means of earning a living. Furthermore, I would use pre-consumer waste as square pieces of bedding fabric to create reversible duvets for my family. I also employed appliqué from fabric scrap to construct and decorate shoe organisers and wall hangings. Many years later, I put a name to the technique: upcycling. Upcycling offers a promising solution to solid waste management and attaining the United Nations Sustainable Development Goals (SDGs). Upcycling transforms disposable items into value-added products (Wegener, 2016), promoting environmental sustainability. Upcycling is one of the most sustainable circular solutions in the waste hierarchy since it typically requires little energy input and can eliminate the need for a new product. Upcycling involves substantial creativity and vision

DOI: 10.4324/9781003380566-21

based on a fundamental environmental consciousness (Sunhilde & Simona, 2017). The products created through upcycling are better than the original ones, unique, and sustainable, and they tell sustainability stories such as 'production with zero waste', 'small is beautiful', and 'start local, but think global' (Wegener, 2016; Earley, 2011; Sunhilde & Simona, 2017). Upcycling reveals new business horizons for small and medium-sized enterprises (SMEs), and companies already established in the labour market, enriching their production (Doble, Böhm, & Porumb, 2021).

My PhD study (2012) was titled 'The Indigenous Dress of the Mau Ogiek People, Nessuit Location, Nakuru County, Kenya'. Upon reflection, the study demonstrates how the community fosters sustainability, particularly zero waste. No item of dress is thrown away. Instead, it is mended and handed over to the next generation, and if it no longer fits, it is enlarged using patchwork. People can learn about sustainable design from indigenous communities' culture. The study led me to present extracts at conferences in Nigeria, Kenya, and Uganda and publish them in scientific journals. Conference presentations include 'Functions of Ethnic Dress of the Mau Ogiek People of Kenya *and* Gender Distinction in the Mau Ogiek People's Indigenous Dress, Kenya' (Njeru, Abong'o, & Okumu, 2012a) in Nigeria in 2011, and 'The Application of Non-material Culture on the Mau Ogiek People's Ethnic Dress, Kenya' (Njeru, 2014a) in Keyna in 2011. In 2012, just before graduating with a PhD, I joined the International Federation of Home Economics (IFHE) in Germany. The year 2012 was the beginning of my publishing endeavours in scientific journals. For instance, the *International Journal for Home Economics* (IJHE) is hosted by IFHE (Njeru, Abong'o, & Okumu, 2012a). The paper demonstrated sustainable fashion aspects such as multifunctional and multi-styled dresses, repair, and hand-me-downs. The concepts aptly resonate with upcycling. Another publication (Njeru, Abong'o, & Okumu, 2012b) focused on cotton, end-of-life (EOL) fibre, thus promoting sustainable fashion. Other journal publications include Njeru, Abong'o, and Okumu (2012c) and Njeru, Abong'o, and Okumu (2012d). The satisfaction that I felt seeing my publications was indescribable. I heard colleagues mostly talking about the difficulty of publishing, and I also questioned my ability but decided to publish anyway.

Publishing extracts from my PhD thesis extended into 2013 with Njeru, Abong'o, and Okumu (2013a), in which we argued that ethnic dress promotes gender equality and equity. Njeru (2013) argues that the continuity and discontinuity of indigenous dress are not particularly negative. The *Global Journal of Human Social Sciences* (GJHSS) editor was impressed by my paper on gender distinctions and invited me to publish in their journal. I accepted the invitation and published Njeru (2014a). In the same year, I presented a conference paper about silk at the Mount Kenya University International Conference. The paper (Njeru, 2014b) is featured in the peer-reviewed conference proceedings. Silk is an EOL fibre, thus promoting eco-fashion. Still, in 2014, the IFHE sponsored me to represent it at the 2nd International Workshop on Health Promotion in Kenya by the Alliance for Health Promotion. I learnt about participatory action research (PAR), an approach amenable to sustainable fashion studies.

In 2015, I encountered Mr Emmanuel Mutungi of Kyambogo University, Uganda, a doctoral student at Maseno University. We met at his research proposal defence at the school level, where I was a member of the post-graduate studies committee. Mr Mutungi invited me to present a paper at the Kampala International Design Conference (KIDeC) and facilitate the pre-conference postgraduate seminar. For the former, the paper was titled 'Types and Construction of Indigenous Dressing of the Mau Ogiek People, Kenya'. For the latter, it was on 'Managing Time and Academic Direction/Focus on a Graduate Programme'.

Nonetheless, the conference papers were never published. It was my first time meeting and engaging with designers worldwide and in different specialisations. I was amazed, impressed, and inspired by their contribution to design. I learnt about sustainable concepts such as distributed renewable energy (DRE), biomimicry, and co-creation. I started building my network in the community of designers. Prof Richie Moalosi, Prof Mugendi M'Rithaa, and I were participants and facilitators at the conference. The same year, I met a sports scientist at a conference in Addis Ababa, Ethiopia, who observed that active sportswear for people living with disabilities (PLWD) is ill-fitting, unsightly, and uncomfortable. He challenged me to design adaptive active sportswear. The scientist's challenge inspired me to submit a research proposal on 'Adaptive apparel for persons living with disabilities' for a methodology work. Unfortunately, the reviewer used abusive language rather than critiquing the proposal. The review was disheartening, but I held on to the concept and later conducted research and published it. The special group of consumers is under-served, neglected, and marginalised by, among others, fashion design practitioners and educators in their collections, teaching, and research. The consumers are forced to wear active sportswear designed for the able-bodied, contrary to sustainable ethos and principles, such as co-design.

In 2016, I received an invitation from Prof Mugendi M'Rithaa to present a paper at the Sustainable Energy for All by Design Conference in Cape Town, South Africa. I read extensively on sustainability and sustainable fashion, researched, and presented a paper (Njeru, 2016). The paper demonstrates how academia is critical in teaching and imparting sustainability concepts to undergraduate and postgraduate students. I also joined the Learning Network on Sustainability (LeNS, www.lenses.polimi.it). I was honoured to meet Prof Carlo Vezzoli of Politecnico di Milano, Design Department, Italy, whom I would refer to as 'the father of Design for Sustainability (DfS)'. His commitment to design for sustainability is unmatched and unwavering and spans decades, to the extent of being awarded an honorary doctorate by the

Federal University of Paraná, Brazil, in 2022. The LeNS conference offered me another invaluable opportunity to engage with designers from around the globe and build up my network. I was taking baby steps in the sustainable fashion *safari*.

Prof Richie Moalosi was impressed by my paper in the LeNS conference proceedings and invited me and his colleague Dr Yaone Rapitsenyane to conduct research in collaboration. We presented a conference paper (Rapitsenyane, Njeru, & Moalosi, 2019a) at the LeNS World Distributed Conference: Designing Sustainability for All, Cape Town, South Africa. From the same research, we published a book chapter (Rapitsenyane, Njeru, & Moalosi, 2019b). From the study, I learnt about sustainable product-service systems (S.PSS), a sustainability concept applied to diverse design fields, including fashion. To my surprise, I discovered among the book's contributors the renowned author, Prof Susan Kaiser, whose book I have used, *The Social Psychology of Clothing: Symbolic Appearances in Context* (2nd ed. revised), for teaching at the University. The same LeNS conference proceedings feature a paper (Njeru & M'Rithaa, 2019) advocating for adopting applied research in sustainability and human-centred approaches in fashion design doctoral theses. The concepts are aligned with sustainable fashion. The conference offered another excellent opportunity to stimulate engagement with designers committed to advancing DfS.

In 2019, I took up the challenge from the earlier sports scientist. I studied 'Junior Sportspersons Living with Physical Disabilities' [Dis]Satisfaction Level with Selected Active Sportswear Attributes: Implications for Sustainable Apparel Design for Social Inclusion in Kenya'. I submitted the manuscript to the IJHE. Nonetheless, it was not accepted for not aligning with the journal's theme. I put the manuscript aside while searching for a relevant journal or conference to submit to. I came face to face with the difficulty of publishing.

The *safari* has faced some challenges, especially the lack of funding for grant proposals. I put together two consortia in 2020 comprising African scholars and applied for funding for 'Empowering Persons Living with Disabilities *and* Specialised Apparel Made from Bamboo for Women Breast Cancer Survivors'. However, they still need to be funded – the projects aimed to enhance the socio-economic inclusion of special needs persons. Bamboo is an EOL fibre, thus sustainable. Then came a surprise invitation from Prof Miguel A. Gardetti and Prof Subramanian S. Muthu (editors of the sustainable textiles and fashion series in Springer Nature) to read our book chapter (Rapitsenyane, Njeru, & Moalosi, 2019b). I took up the challenge and submitted the manuscript that IFHE rejected. Initially, the reviewers did not accept the paper for lack of a rigorous scientific approach. I took the feedback positively, beefed up the paper, and it was published in another series by the same editors (Njeru, 2021). Sustainable fashion promotes the social inclusion of all end-users, including PLWD.

In late 2021, during the COVID-19 pandemic, Prof Moalosi invited me to submit a book chapter for *Indigenous Technology Knowledge Systems: Decolonising the Technology Education Curriculum*, edited by Prof Mishack Gumbo and Prof John Williams. My paper discusses the 'Nexus of Indigenous Technological Knowledge Systems and Design Education in Afrika's Higher Education Institutions'. The book tackles sustainability issues, and was published in 2023.

The year 2022 saw an upsurge in scholarship after the lull occasioned by the COVID-19 pandemic. Early in the year, Prof Moalosi invited me to join the International Upcycling Research Network hosted by De Montfort University (UK) and the University of Botswana, with global membership. A series of seminars were held between June 2022 and January 2023. Prof Moalosi and I presented a paper about 'Kenyan Fashion Designers' Knowledge, Attitude, and Practice of Upcycling'. On upcycling, Kenya's fashion practitioners are knowledgeable, have a positive attitude, and practise it extensively. I accepted an invitation from one of the members, Elizabeth Burton, to collaborate in a workshop slated for September 2023 for textile colouration using *shibori* – a Japanese resist-dyeing technique. The project also incorporates upcycling and will involve undergraduate fashion design students at Kirinyaga University. To my surprise and awe, Prof Rebecca Earley, whose publications I have cited in some papers, is a member of the network.

Prof Moalosi also invited me to join a team of African designers to contribute a chapter to a book he is editing with Dr Yaone Rapitsenyane. The book's title is *African Industrial Design Practice: Perspectives on Ubuntu Philosophy* (Routledge). With Ms Nani Setlhatlhanyo and Mr Polokano Sekonopo, we researched and submitted our chapter on 'The Business Case of Design in Afrika'. *Ubuntu* is an African philosophy aligned with sustainability. In June, Prof John Williams shared the call for the 11th Biennial International Design and Technologies Teaching and Research Conference (DATTArc) at Southern Cross University, Australia. I took it up and presented a paper on 'Antecedents to the Non-adoption of Fashion Design for Sustainability (FDfS) and Applied Research to my Master's Thesis', focused on Kenya. The paper was well received for its contribution to postgraduate research and sustainability, and is due for publication soon as part of the conference proceedings. In the same year, Prof. Sabu Thomas, Vice Chancellor of Mahatma Gandhi University, India, invited me to be a speaker at their International Online Conference on Reuse, Recycling, Upcycling, Sustainable Waste Management, and Circular Economy (ICRSC – 2022). He was highly impressed with my publications. My presentation was about 'Situating Fashion Design Actors at the Core of Design for Sustainability (DfS) Discourse'. The paper highlighted the hits and gaps concerning sustainability among fashion actors: academia, practitioners, government, and community, and proposed a way forward to enhance their participation. I also accepted a request to chair one session during the conference. Acting on Prof. Thomas's request to invite other scholars, mainly from Africa, to present at the conference, I invited two, but only Prof. Moalosi responded. He gave an inspiring plenary presentation detailing the International Upcycling Research Network project and encouraged conference participants to join.

Sustainable fashion *safari* in teaching

It is paramount to educate the next generation of designers in the first years of design education in all sustainability fields: environmental, social, and economical, and in subjects such as upcycling, zero-waste design, and disassembly so that they can grow with the idea of designing not just one generation of products (Vezzoli et al., 2022). Furthermore, it is paramount to teach sustainability and the impact of the fashion sector and instil skills which enable the students to pursue fashion design as a sustainable practice (Grose, 2013). Raike et al. (2009) advocate for engaging students in real-life projects with actual end-users. Higher education institutions are challenged to turn and fully integrate sustainability principles into practice through, *among other things*, curricula and the teaching of new competencies (Ferreira, Souleles, & Savva, 2019).

On curriculum, in 2017, inspired by new-found knowledge of sustainability, I revised the Diploma and Bachelor of Science in Fashion Design and Marketing at Machakos University, Kenya. I incorporated a unit on Sustainable and Ethical Fashion Design. I pioneered its delivery as studio-based, as well as inducting the diploma tutors on the same. The undergraduate students' projects comprised upcycled, transformable, multi-styled, and reversible apparel. All the creations relate to sustainable fashion. While serving as the chair of a department at the same university, I sought pre-consumer textile waste from the Export Processing Zone at Athi River. Leading international fashion brands produce their products in Kenya, especially from denim and khaki fabrics. The students were complaining about the high cost of project materials. Using textile waste reduced the students' costs of buying new materials. Unfortunately, I was informed that all the waste, including textiles, is incinerated. The practice is contrary to Doble, Böhm, and Porumb (2021), who stress that in upcycling, collaborations must be established between large-series producers and the small business environment, comprising small workshops, non-profit organisations, or school organisations, which can reuse them.

When I joined Kirinyaga University in 2021, I was tasked with revising the BSc and develop a new MSc in Fashion Design and Textile Technology programme. My proposal for incorporating a sustainable fashion unit was disregarded for both programmes. However, in every studio-based unit I teach, I stress students to co-design with actual end-users and create good-quality products. I also encourage them to upcycle second-hand fabrics and soft furnishings for their projects to reduce costs. These initiatives foster net zero. Teaching resumed during the COVID-19 pandemic via online delivery. In June–September 2021 and March–June 2022, I taught the unit Fashion Design for Special Groups. This unit has been predominantly delivered as theory. I took a studio-based approach incorporating co-design and Design Thinking. The students collaborated with actual end-users, special needs persons (SNP), and people exposed to occupational hazards (PEOH) to design and construct specialised apparel and practical/work clothes, respectively.

The outcome was highly beneficial to both the students and the end-users. Co-creation leads to near-perfect fit and attachment to the fashion products, thus enhancing longevity/durability and delayed disposal. The positive outcome was shared in a paper titled 'Fashion-abled Special-needs Groups: Implications of Co-creation for Fashion Design Undergraduate Degree Pedagogy and Sustainability', in the University of Nairobi's 4th Annual International Conference on Research and Innovation in Education (AICRIE) (2022), Kenya. One conference participant complimented me on the presentation via email. The main challenge in adopting the co-creation approach was convincing and guiding the students about interacting with actual end-users and designing and producing complete apparel for the first time in the course. However, with proper instructions, guidance, encouragement, and class presentations throughout the semester, the students could complete the projects to the end-user's satisfaction. I tweaked and submitted the paper to *CoDesign: International Journal of the Arts and Design*. I am awaiting the reviewer's feedback. Co-creation or co-design is a process in which new solutions are suggested and new meanings are created by diffuse design performed by everybody and expert design performed by trained designers (Hu et al., 2019).

The IFHE annually celebrates World Home Economics Day on 21 March. This year's (2023) theme was *Waste Reduction Literacy*. I shared the celebration with the undergraduate fashion design students, and we brainstormed on fashion's unsustainable production and consumption and how to address the IFHE theme. I inspired the students using France24 television clips on fashion, local fashion designers, and the link for the International Upcycling Network. We agreed to construct fashion products aligned with the theme and textile waste and curate a fashion show and exhibition. We planned to invite fashion industry captains and secondary school students, among other guests, to market the design students and the programme. However, the university management declined our event host request. The news discouraged the students. Eight out of 133 students produced apparel, apparel accessories (bags), soft furnishings (floor mat and organiser), and interior accessories (flower vases, storage baskets and bags, and paintings) from pre- and post-consumer waste, as shown in Figure 15.1.

FIGURE 15.1 Upcycled pair of pants made from *bandana* and white scrap fabrics (left) and upcycled dress fashioned from post-consumer blouses (100% cotton) and pre-consumer organza (right)

The students could express their creativity in the projects and foster sustainable design. The students' opinion about the project was positive: 'Designers can make great things out of waste fabrics from their projects. When given a thought, those small scraps of fabric will not disappoint in making whatever kind of garment a designer wants. Upcycling is cheap and profitable. My model loved the patchwork and went the extra mile of buying it to use it for a modelling competition as streetwear'; 'Recreating new outfits from old clothes is a good and easy way of reducing waste in the fashion industry. One can make bags, jewellery, headwraps, crop tops, shorts, and many other items from old outfits that one is not using instead of disposing of them'; 'I greatly enjoyed taking part in this project. It gave me a cheaper alternative to draw my paintings on since the canvas fabric used for painting is expensive and can be limiting at times'; 'This project can help reduce the textile product waste disposed to the environment'. The outcome of the project was compiled in a manuscript titled *Textile Solid Waste Management: Adding Value and Uniqueness through Design*, which I submitted to the Annals of the University of Oradea Fascicle of Textiles, Leatherwork, Romania. I am awaiting the reviewer's feedback. Designing upcycled products requires a new radical cultural mindset change (Holtström, Bjellerup, & Eriksson, 2019) from the culture of throwing away to a design approach of component modularity that allows products to be deconstructed and reconstructed. Furthermore, it demands innovation and a futuristic creativity approach (Johnson & Plepys, 2021).

In late 2022, Ms Ryna Cilliers, faculty at Cape Peninsula University of Technology (CPUT), South Africa, approached me via LinkedIn about examining a Master of Technology (M.Tech) thesis. I readily accepted. The appointment was communicated in February 2023. The thesis designed an academic gown for wheelchair users, issues aligned with my research interests and sustainable fashion. A significant surprise and fulfilment is that in February 2023, Kirinyaga University procured numerous textbooks whose titles I had submitted in May 2021. Almost all the textbooks (hard copy) are sustainability-oriented from leading sustainability authors such as Kate Fletcher, Carlo Vezzoli, Miguel Angel Gardetti, Fabrizio Ceschin, and Alison Gwilt, the majority of whom I have interacted with. I am supervising three PhD and one master's candidate at different universities. The PhD candidates are engaged in sustainability concepts: upcycling for artworks, apparel collection for plus-size millennial women, and social entrepreneurship and community development. Fashion design practitioners, educators, and researchers have neglected plus-size women, leading to unsustainable consumption of apparel. The master's candidate's research is on life cycle design.

Sustainable fashion practice and future initiatives

Alongside teaching and scholarship, I also practise designing and constructing apparel, fashion accessories, and soft furnishings anchored in cultural sustainability – specifically, a cap made from the popular Masaai *shuka*. The inspiration was from African cloaks/caps, common in diverse ethnic dresses. Other creations comprise a laptop bag made from used bedlinen, upcycling, a shopping bag from waste Masaai *shuka*, upcycling, a dress from *leso/khanga*, a 100% cotton cloth indigenous to the Swahili people, a fabric that was too stiff to be used for apparel which, instead of discarding it, was used to construct a pair of curtains, and reversible duvet, upcycling.

My concept notes titled 'Mainstreaming Sustainable Fashion in Technical and Vocational Education and Training Curriculum for Persons Living with Disabilities in Kenya', submitted to Harvey Mudd College's workshop, was accepted for presentation in June 2023. Nevertheless, I withdrew from attending the conference due to financial constraints. I intend to conduct research and develop a manuscript on sustainable share, which I will share with the institution.

Conclusion

My sustainable fashion *safari* has had its fair share of hits and misses, but the former outweigh the latter. The *safari* involves pedagogy, research, publishing, and conference/seminar presentations. The *safari* is a learning curve about sustainability theories, models, research methods, pedagogy, publishing, and grant writing. I have learnt the art of collaborative research. Fulfilment emanates from knowing that I am participating in the global discourse concerning sustainable fashion and in creating a pool of future fashion design professionals who are responsive and empathetic. Given opportunities and proper instructions in class projects and design-related activities, fashion design students can comprehend and meaningfully incorporate sustainability principles and ethos in their projects. The challenges the designers encounter in the projects offer possibilities for in-depth learning and acquiring design-related skills. I am open to more engagements and collaborations.

References

Doble, L., Böhm, G., & Porumb, C. L. (2021). Smart valorification of recyclable textile waste. *Annals of the University of Oradea Fascicle of Textiles, Leatherwork*, 22(1), 33–36. http://textile.webhost.uoradea.ro/Annals/Volumes.html.

Earley, R. (2011). *Upcycling textiles: Adding value through design*. Paper presented at the KEA's Towards Sustainability in the Fashion and Textiles Industry. http://ualresearchonline.arts.ac.uk/4023/.

Ferreira, A. M., Souleles, N., & Savva, S. (2019). *Upscaling local and national experiences on education for social design and sustainability for all to a wider international arena: considerations and challenges*. The LeNS World Distributed Conference: Designing Sustainability for All. www.lensconference3.org.

Grose, L. (2013). Fashion design education for sustainability practice: Reflections on undergraduate level teaching. In M. A. Gardetti, & A. L. Torres (Eds), *Sustainability in Fashion and Textiles: Values, Design, Production, and Consumption*. Routledge, pp. 134–147.

Holtström, J., Bjellerup, C., & Eriksson, J. (2019). Business model development for sustainable apparel consumption: The case of Houdini Sportswear, *Journal of Strategy and Management*, 12(4), 481–504. https://doi.org/10.1108/JSMA-01-2019-0015.

Hu, H., Bai, F., Hao, D., & Zhou, J. (2019). *Research of sustainable product service systems on Chinese minority brand context*The LeNS World Distributed Conference: Designing Sustainability for All. www.lensconference3.org.

Johnson, E., & Plepys, A. (2021). Product-service systems and sustainability: Analysing the environmental impacts of rental clothing. *Sustainability*, 13, 2118. https://doi.org/10.3390/su13042118.

Njeru, S. (2013). Continuity and discontinuity of the indigenous dress of the Mau Ogiek People, Nessuit location, Nakuru county, Kenya. *International Journal of Art and Art History*, 1(1), 10–21. www.aripd.org/ijaah.

Njeru, S. (2014a). The application of non-material culture on the Mau Ogiek People's ethnic dress, Kenya. *Global Journal of Human Social Sciences*, 14(1), 91–100. www.GlobalJournals.org.

Njeru, S. (2014b). *Kenya–China co-operation in the textile and clothing industry through silk production and the resultant socio-economic empowerment of Kenyans.* First International Research and Innovation Conference. Nairobi, Kenya, 28–30 August, pp. 18–24.

Njeru, S. (2016). Incorporation of sustainability into fashion design degree programmes in Kenya. *Sustainable Energy for All by Design,* 393–404. www.lenses.polimi.it.

Njeru, S. N. (2021). Junior sportspersons living with physical disabilities' [dis]satisfaction level with selected active sportswear attributes: Implications for sustainable apparel design for social inclusion in Kenya. In S. S. Muthu, & M. A. Gardetti (Eds), *Sustainable Design in Textiles and Fashion.* Springer Nature Singapore, pp. 53–84. https://doi.org/10.1007/978-981-16-2466-7.

Njeru, S., Abong'o, S., & Okumu, C. (2012a). Functions of ethnic dress of the Mau Ogiek People of Kenya. *International Journal of Home Economics,* 5(2), 227–244. www.ijhe.org.

Njeru, S., Abong'o, S., & Okumu, C. (2012b). The creative use of cotton in fashioning the Mau Ogiek People's ethnic dress – Kenya. *Africa – Dynamics Social Science Research,* 2(1), 118–127. www.aceser.net.

Njeru, S., Abong'o, S., & Okumu, C. (2012c). Universality and diversity of the Mau Ogiek People's ethnic dress, Kenya. *Journal of Alternative Perspectives in the Social Sciences,* 4(2), 465–484. www.japss.org.

Njeru, S., Abong'o, S., & Okumu, C. (2012d). Forces that occasioned the Mau Ogiek People's ethnic dress, Kenya. *International Journal of Humanities and Social Science,* 2(13), 192–197. www.ijhssnet.com.

Njeru, S., Abong'o, S., & Okumu, C. (2013). Gender distinctions in the Mau Ogiek People's indigenous dress, Kenya. *International Journal of Education and Research,* 1(3), 103–118. www.ijern.com.

Njeru, S., & M'Rithaa, M. (2019). *Fashion design-related doctoral studies in selected Kenyan Universities: Advancing applied research in sustainability and human-centred approaches.* The LeNS World Distributed Conference: Designing Sustainability for All. www.lensconference3.org.

Raike, A., Botero, A., Dean, P., Iltanen-Tähkävuori, S., Jacobson, S., Salgado, M., & Ylirisku, S. (2009). Promoting design for all in Finland. Experiences of successful collaboration of users and designers at the University of Art and Design Helsinki. *Design For All Institute of India,* 4(9), 61–94.

Rapitsenyane, Y., Njeru, S., & Moalosi, R. (2019a). Sustainable product-service systems: A new approach to sustainable fashion. The LeNS World Distributed Conference: Designing Sustainability for All. www.lensconference3.org.

Rapitsenyane, Y., Njeru, S., & Moalosi, R. (2019b). Challenges facing the fashion industry in implementing sustainable product service systems in Botswana and Kenya. In A. Gwilt, A. Payne, & E. A. Ruthschilling (Eds), *Global Perspectives on Sustainable Fashion.* Bloomsbury Visual Arts, pp. 236–245. www.bloomsbury.com.

Sunhilde. C., & Simona, T. (2017). Can 'upcycling' give Romanian's fashion industry an impulse? *Annals of the University of Oradea Fascicle of Textiles, Leatherwork,* 18(1), 187–192. Available: http://textile.webhost.uoradea.ro/Annals/Volumes.html.

Vezzoli, C., Conti, G. M., Macrì, L., & Motta, M. (2022). *Designing Sustainable Clothing Systems: The Design for Environmentally Sustainable Textile Clothes and Its Product-Service Systems.* FrancoAngeli.

Wegener, C. (2016). Upcycling. In V. P. Glaveanu, L. Tanggaard, & C. Wegener (Eds), *Creativity – A New Vocabulary.* Palgrave MacMillan.

16

SUSTAINABILITY IN THE CREATIVE ECONOMY

Experience in the textiles and fashion industry

Walter Chipambwa

The textile manufacturing experience

I was born and raised in Zimbabwe, a country traditionally known for one of the finest cotton grades in Africa. I did my first degree (Bachelor of Engineering) in Textile Technology at a local university, which opened my mind to the world of textile manufacturing and apparel design. I enrolled for my first degree in 2003, the same period the manufacturing sector was going through a difficult time, with many companies facing operational difficulties. The textile sector in Zimbabwe faced numerous challenges that included high production costs, power shortages, the threat of cheap imports, and an economic slowdown that led to the rapid de-industrialisation of the sector, resulting in the eventual closure of some companies (Mvula & Chummun, 2020).

Despite these challenges, some companies managed to survive the hardships. Such companies mainly relied on importing raw artificial fibre and exporting it to other regional nations. As part of the final year of my studies, I was exposed to the world of colouration technologies through my dissertation, as it focused on using natural dyes instead of the harmful synthetic dyes that result in effluent discharges that affect aquatic life. This project shaped my career, as my interests were mainly in colour physics and how colour affects design interpretation in life. With natural dyes, I saw an opportunity to apply colour at the lowest cost, as I could relate to various colours one could extract from nature. This undergraduate project experience opened my eyes to the other dyeing parameters and the business side of the textile industry, as the extraction and expansion of the natural dye colour gamut has its fair share of challenges (Chipambwa & Nyathi, 2017). My first job in the textile industry in 2007 was to manage the wet process (dyeing) in an acrylic dyeing setup. Dyeing is one of the most energy-intensive and polluting stages

DOI: 10.4324/9781003380566-22

of textile production, as it involves heating large amounts of water, using synthetic chemicals, and generating wastewater. Sadowski, Perkins, and McGarvey (2021) report that textile processes, namely, weaving, dyeing, and finishing, annually account for approximately two percent of greenhouse gas emissions worldwide. As a graduate trainee, I led a team to push for the International Organization for Standardization (ISO) certification. The company wanted to expand its export base by complying with regional quality and environmental standards.

As the company sought to be ISO certified, it had to implement specific methods in effluent treatment to reduce its overall carbon footprint of the dyeing process. The quality circles team I had led identified the dyeing process as the first project requiring an immediate response due to its effluent discharge. The company had to adopt environmental standards set out by the Environmental Management Agency (EMA) and general quality standards as stated by the Standards Association of Zimbabwe (SAZ). This resulted in the company constructing effluent treatment tanks and a water recycling plant, as the dyeing process consumes a lot of water. The dyeing process of acrylic fibre happens at very high temperatures of 110–140°C in the pH range of 4.5–5, usually controlled by the addition of acetic acid. To achieve this, high-temperature steam from the coal-fired boilers was used to heat the dye vessels holding large amounts of water, usually 9,000 litres. The effluent from the dyeing process is difficult to analyse, as the chemical composition of the toxic elements can be challenging to detect. Even though synthetic dyeing produces long-lasting colours that are very bright and of a broader range, unlike natural dyes, the company could not substitute these dyes. However, they were known to harm the environment and pose a hazard to aquatic life. As a developing economy with enormous coal resources, the issue of using steam boilers was not taken as an area that could be addressed to reduce greenhouse gas emissions despite the dangers this posed to the employees and the environment at large. Projects to start treating the effluent and recycling the wastewater were designed and implemented following the recommendations from EMA despite the limited financial resources in testing the effluent particles. This process of testing pH and sending results to the environmental agency had its fair share of challenges, as it was an expensive process that required sending samples to a specialist lab.

In most cases, results would come after the water had already been discharged into the stream behind the factory. When the pH was too acidic, sodium sulphate was added to the effluent to neutralise the acidic waste. The company also sponsored my master's studies (Goriwondo et al., 2012), where the overall objective of the study was to devise a system that could schedule dyeing jobs with the aim of saving changeover times from one colour to another and also reduce the amount of water used per cycle in cleaning the machine after a dyeing cycle. The vast quantities of water and chemicals cleaning the machine after one dyeing cycle motivated the project. The objective was to come up with a scheduling of dyeing jobs on the four dyeing machines such that there is no machine cleaning when changing colours related to each other. For example, dyeing colours pink, maroon, and scarlet would be done on the same machine, so there was no cleaning time. The project was implemented in 2011 at the company and resulted in huge savings in the amount of water used daily as well as the cleaning chemicals. In 2007, I also proposed adopting a quicker dyeing method that resulted in reduced energy consumption in the company, by substituting dispersed dyes that required more processing time with the basic dyes that required less time whilst achieving excellent, bright colours. The company also made acrylic tufted carpets and dyed the fibre according to the specific order colour. Due to the sporadic power outages experienced from 2007 to 2011, the company had to opt for new product ranges with a mix of colours that proved popular among clients. These colours resulted from mixing different wrongly dyed fibres but expertly woven into new designs that gave new design pathways for various hotels.

The academic journey

In 2013, I left the manufacturing industry to join academia as I liked the thought of exploring new challenges. Through my master's studies, I developed new interests in improving production efficiency through computer-aided design and systems design. In my first year of teaching at the university, I taught computer-aided design to students studying for a fashion design degree. I blended the knowledge from my internship experience at a clothing company in 2007 with the experience I had attained in a textile manufacturing company for six years. The concept of less is more became my new interest as I explored ways of using the computer to design in a minimalistic way. In the conference paper, Chipambwa and Samwanda's (2016) study sought to establish the role played by designers in promoting a green economy. The argument is that they are the first people who can communicate certain information to the public that can promote sustainable behaviour or practices. The study was focused on how these designers position themselves in a sustainable design circle through their actions and inputs to clients to promote sustainability. In the same year (2016), I enrolled for my PhD studies with the University of Botswana to further my knowledge of sustainable design in small and medium-sized enterprises (SMEs). SMEs have become engines of growth for many nations, and as such, the study sought to unpack sustainable design-driven innovation in this crucial sector. Going ahead, I started developing my academic presence through my first book chapter, which was on designing for functionality (Chipambwa, 2017), exploring the issues of aesthetics, cost, and functionality in design in general. The chapter was inspired by the need to keep the design simple, smart, and straight to the point through the *keep it simple, stupid* (KISS) concept.

Upcycling as a sustainable practice

With the new developments in the clothing manufacturing sector, where fast fashion has been taking centre stage, our study (Chuma et al., 2018) explored how clothing manufacturing companies could adapt to this era from the developing economy context. The fast fashion era has contributed to vast amounts of clothing waste found in many landfills worldwide and to the growth of trade in second-hand clothes (SHC). This trade in second-hand clothes is evident in most African countries and has affected the growth of the local clothing industries. Issues of health risks associated with second-hand undergarments were also reported in the study by Chipambwa, Sithole, and Chisosa (2016). Despite these health risks, trade in such SHC undergarments is still widespread, as fashion brands, quality, size, and price issues have become the key selling points for such products. This has also contributed to the collapse of the local clothing companies in Zimbabwe, thus leaving a huge gap that is difficult to fill, and this was also further exacerbated by the economic woes that the general populace faced as the economy went down on its knees. Due to the different sizing systems used in the countries of origin of these SHCs, they now pose another problem: they were discarded in these developing countries as they cannot fit the local people.

Some designers have gone to the extent of upcycling these garments as they are made from good quality fabric deemed fashionable. In 2019, we published a journal paper on a study done with fashion design students on how they carried out the upcycling process (Chuma et al., 2019). The study revealed that the upcycling process took longer for one to produce a garment as each process produced a unique item that could be difficult to imitate. Upcycling the carbon footprint of the clothing sector can be reduced as the material is reused, unlike in fast fashion. The major problem cited by the designers was the pricing of the product from upcycling, as customers consider its inferior quality. Also, the customers might develop a negative attitude towards the product as they prefer something made from new fabric. These were cited as the major problems associated with upcycling, thus making it difficult for the designers to convince their customers to purchase. To promote the concept of upcycling, the customers must first be made aware of their role in reducing the carbon footprint. Once they appreciate their role, customers can push the manufacturing sector to also play its role in promoting a greener future.

Upcycling for community empowerment

In 2020, we started a project to empower rural women through upcycling of waste from a clothing factory. The waste management methods implemented by the clothing companies inspired the project. These rural women are usually idle during the day when it is not the farming season. To supplement their income, the upcycling project was introduced through the Musor-omuchena Life Skills Centre in the rural district of Makonde in Mashonaland West Province in Zimbabwe. The skills centre invited women to come and engage in skills development during their free time, and they were given various fabric wastes or offcuts to use on their designs. These women designed and made various products, including bags, blankets, skirts, ropes, and mats. Some products were sold through an open-day market fair, and the women used some in their homes. The ongoing project was presented at an online seminar on upcycling in 2022 (Sung, 2022). The project exposed me to different perspectives on how waste material can be converted into something more useful and of higher value. It resulted in another project on upcycling wooden floor tiles to make accent furniture pieces that is also ongoing, as the tiles were obtained from a hotel doing a floor makeover, removing wooden tiles and replacing them with ceramic tiles. Despite its size, these projects have proved that upcycling can be instrumental in promoting circularity in an economy. As an interdisciplinary designer, I have embarked on a journey to promote sustainable behaviour among designers as they are the key players in whatever product they design for the consumers, be it fashion design, product design, or graphic design.

Digital fashion designing for sustainable futures

Traditionally, the process of fashion designing involves manually drawn sketches using pencil on paper. The developments in the world of computers have resulted in software packages that allow one to use the computer to generate the same images using the computer, thus no need for pencil and paper. As one of the courses I taught from my first day teaching at university, I have encountered many applications that compete in executing similar tasks. Through digital designing, the need for physical clothing samples is reduced. This results in considerable savings in materials, energy, water, and emissions associated with producing and transporting physical products. In computer-aided design for fashion, the critical aspect is removing the need for manual methods in designing, illustration, and pattern making. In one of my studies (Chipambwa & Chimanga, 2018), we sought to explore the key issues affecting the design industry, specifically focusing on software. Access to fully licensed software versions is commonly challenging in a developing economy. This has affected the uptake and integration of technology within clothing manufacturing setups despite the benefits of investing in computer-aided design for fashion. The COVID-19 pandemic brought many changes concerning technology adoption in the clothing industry. Life had to go on, and designers had to work remotely and share new trends and new fashion ideas, thus promoting the digitalisation of the fashion sector. This digital transformation led to a more paperless society in all manufacturing industries worldwide. The digitalisation of the clothing sector also promotes sustainability as manufacturing costs are reduced.

Technologies such as artificial intelligence (AI), virtual reality (VR), and augmented reality (AR) have become more powerful tools that are promoting sustainable business models in the fashion sector (Chipambwa et al., 2022). The work on virtual exhibitions explored how final-year students could hold their events virtually to reach a broader audience to showcase their design ideas for the final-year fashion design project (Chipambwa et al., 2022). Through virtual exhibitions, the need for resources to physically put together an event is removed and saves on materials, thus positively promoting a net zero fashion industry. At the beginning of 2023, we also embarked on another project to promote virtual reality and the metaverse, where fashion designers and industrial designers are expected to develop new products or ideas that can be launched through a metaverse channel in the first quarter of 2024. The main idea is to promote the physical and digital presence in the design world, now called *phygital*. Digital fashion promotes creativity and innovation among designers as they can quickly develop new sustainable solutions for the sector. The computer-aided design also promotes efficient use of resources in pattern making as efficient pattern markers are produced for each pattern, thereby reducing waste and errors in the pattern-making process.

Sustainable behaviour of small and medium-sized enterprises

My PhD journey has not been as smooth as I might have liked, and I have had many challenges, including funding and the COVID-19 pandemic. Nevertheless, my study on small and medium-sized enterprises (SMEs) gave me insights into the sustainable design behaviour of furniture manufacturing SMEs in Zimbabwe. A significant issue of the study was the lack of a proper framework or policy that can promote sustainable behaviour among manufacturing SMEs. In the article by Chipambwa et al. (2023), we highlight the need to educate SMEs on the benefits of circularity in manufacturing and how sustainability can be used as leverage for a better customer–supplier relationship. The study also discussed the role of designers in promoting sustainability, as their creative design ideas are the key selling points of furniture products. Generally, SMEs in developing economies lack policies that compel them to adhere to sustainable behaviour, or if there is a policy in place, there is a lack of proper enforcement. This can be attributed to several factors ranging from limited financial resources to invest in low-carbon technologies, lack of knowledge on their effect on the environment, and lack of effort to promote a sustainable drive in their manufacturing process. The study on SMEs also concludes that there is a need to develop more environmental policies or incentives for SMEs and large firms so that they align their net zero goals and actions.

Can developing countries adopt the net zero concept?

My experience has taught me that a net zero economy can be achieved in developing economies with significant support and collaboration from developed economies. One of the challenges in attaining net zero in developing economies is their overdependence on fossil fuels for energy. There is a need to build skills and capacities for net zero in developing economies through education, training, awareness-raising, and knowledge-sharing. Adoption of net zero in developing economies is possible as long as there are concerted efforts from all the stakeholders to achieve and drive towards the same goal. In my experience in the textile and clothing industry, many processes can be earmarked as a priority to reduce carbon emissions, including dyeing, printing, weaving, spinning, pattern making, and designing.

References

Chipambwa, W. (2017). Designing for functionality. *Issues in Art and Design* (Muzenda & Chivhanga ed.). Chinhoyi University of Technology Publications, 63–76.

Chipambwa, W., & Chimanga, T. (2018). Creative design software: Challenges and opportunities to the graphic designer in Zimbabwe. *Journal of Graphic Engineering and Design*, 9 (1), 29–35. doi:10.24867/jged-2018-1-029.

Chipambwa, W., Moalosi, R., Rapitsenyane, Y., & Molwane, O.B. (2023). Sustainable design orientation in furniture-manufacturing SMEs in Zimbabwe. *Sustainability*, 15 (9), 7515. https://doi.org/10.3390/su15097515.

Chipambwa, W., Munyaka, C., & Mutungwe, E. (2022). Virtual exhibitions for final capstone fashion design project: A case of the Chinhoyi University of Technology final year students. *International Journal of Costume and Fashion*, 22 (1), 25–35.

Chipambwa, W., & Nyathi, J.A. (2017). Expansion of natural dye colour gamut for use by the local textile and craft industry. *Current Trends in Fashion Technology & Textile Engineering*, 1 (2), 29–34.

Chipambwa, W., & Samwanda, B. (2016). *Towards a sustainable green economy: The role of graphic designers in Zimbabwe*. The 1st Institute of Lifelong Learning and Development Studies International Research Conference, Chinhoyi University of Technology, School of Art and Design, Zimbabwe. https://www.researchgate. net/publication/336369233_The_1_st_Institute_of_Lifelong_Learning_and_Develop ment_Studies_International_Research_Conference_Interfacing_Technology_Langua ge_and_Knowledge_Management_with_Sustainable_Development_BOOK_OF_ ABSTRACT.

Chipambwa, W., Sithole, L., & Chisosa, D.F. (2016) Consumer perceptions towards second-hand undergarments in Zimbabwe: A case of Harare urban dwellers. *International Journal of Fashion Design, Technology and Education*, 9 (3), 176–182. doi:10.1080/17543266.2016.1151555.

Chuma, C., Chipambwa, W., & Komichi, R. (2018). Staying competitive in the fast-fashion era in a developing economy. *International Journal of Costume and Fashion*, 18 (2), 1–12. doi:10.7233/ijcf.2018.18.2.001.

Chuma, C., Muza, T., & Chipambwa, W. (2019). An examination of the upcycling design process of used clothes among fashion design students: Towards an approach for sustainable fashion design practice. *International Journal of Costume and Fashion*, 19 (2), 59–69. doi:10.7233/ijcf.2019.19.2.059.

Goriwondo, W.M., Chipambwa, W., & Mhlanga, S. (2012). *Product scheduling in a multi-product colour processing facility. Case Study at TN Textiles (Pvt) Ltd*. Proceedings of the 2012 International Conference on Industrial Engineering and Operations Management, Istanbul, Turkey, 3–6 July.

Mvula, D., & Chummun, B.Z. (2020). Investigating industrial upgrading and business sustainability in the textile manufacturing industry in Zimbabwe. *Journal of Contemporary Management*, 17(2), 445–471. https://orcid.org/0000-0001-9043-5339.

Sadowski, M., Perkins, L., & McGarvey, E. (2021). Roadmap to net zero: Delivering science-based targets in the apparel sector. Working paper by World Resources Institute: Washington, DC, USA. https://doi.org/1 0.46830/wriwp.20.00004.

Sung, K. (2022). Upcycling in the community: Towards an inclusive, sustainable cycle. https://www.youtube.com/watch?v=q1DnX-fRXV4&t=574s.

CONCLUSION

Progress towards a net zero carbon society

Richie Moalosi, Kyungeun Sung and Patrick Isherwood

This book, *Research Journeys to Net Zero: Current and Future Leaders*, is highly relevant and timely in a world facing an increasingly urgent climate crisis. It acts as a roadmap for many industries and sectors to transition to a net zero economy, including research, engineering, design, and innovation, as well as the energy, transportation, fashion, and service industries. Students, academics, policymakers, and practitioners can all benefit and obtain insight from the material presented here. The net zero transition proposed is equitable, promotes socio-ecological sustainability, and is aligned with broader sustainable development goals. The arguments raised in this book aim to ensure the robustness of net zero as a framework for climate action. The book's arguments are based on the seven attributes of net zero as a frame of reference, thus (i) front-loaded emission reductions, (ii) comprehensive emission reductions, (iii) cautious use of carbon dioxide removal, (iv) effective regulation of carbon offsets, (v) equitable transition to net zero, (vi) socio-ecological sustainability, and (vii) new economic opportunities (Fankhauser et al., 2022). Readers will find the book valuable for the following reasons.

Guiding sustainable practices: The book guides various sectors, industries, and individuals on adopting sustainable practices in three domains of sustainable development: environmental, social, and economic, including sustainable technologies, innovation, and methodologies. It aims to help entities focus on sustainable and cleaner production by adopting digitalisation to unlock the potential for achieving a circular economy and, ultimately, net zero emissions (Agrawal et al., 2023). The book contributes to understanding the benefits of adopting sustainable entrepreneurial practices, and helping entities reformulate more appropriate and practical strategies that promote continuous sustainable innovation to reduce their environmental footprint (Qader et al., 2022; Franco & Rodrigues, 2019).

DOI: 10.4324/9781003380566-23

Urgent need for climate action: One of the most critical global concerns is the need to resist climate change and move toward a net zero carbon future, which is emphasised throughout the book. It emphasises the critical role of research and development in advancing practical solutions. It pushes industry away from the existing linear take–make–waste paradigm and toward circular business structures and practices (Sadowski et al., 2021).

Informing policy and decision-making: The book's insights will help educate decision-makers, government agencies, and establishments regarding practical methods for achieving net zero carbon emissions. It serves as a knowledge foundation to aid in the creation of evidence-based policies by decision-makers.

Raising public awareness: The book educates readers about the adverse effects of human activity on the environment and the necessity of switching to sustainable methods. It equips people with the knowledge they need to make decisions that support sustainable development and a sustainable future.

Fostering interdisciplinary and transdisciplinary collaboration: By combining knowledge and expertise from various fields – science, engineering, design, innovation, transportation, fashion, and service sectors – this book fosters interdisciplinary and transdisciplinary collaboration. This transdisciplinary approach seeks to dissolve disciplinary silos and is essential for generating comprehensive solutions to complex and apparently intractable sustainability concerns (Kitts et al., 2011). Such an approach may assist in achieving the United Nations Sustainable Development Goals.

Showcasing best practices and case studies: The book inspires and stimulates readers by showcasing successful initiatives and projects that have made considerable progress toward attaining net zero emissions. Valuable lessons can be drawn from such studies and generate a positive impact.

Empowering future generations: Academics, students, and young professionals are encouraged to concentrate on environmentally friendly research and innovation. This book aims to inspire future generations to take the initiative in addressing environmental problems and to bring about constructive social change.

Global impact and collaboration: The book promotes a global viewpoint and collaboration by emphasising worldwide initiatives and effective models from many locations across different continents. It proves that achieving net zero emissions is a community effort that calls for worldwide collaboration and cooperation. The saying "If you want to go fast, go alone; if you want to go far, go together" is an African adage aligned with this initiative. The proverb strongly emphasises the value of collaboration and teamwork in achieving long-term success.

The book chapters for *Research Journeys to Net Zero: Current and Future Leaders* were based on the following critical areas and sectors of the economy, forming the five parts of the book:

Part I Science and Engineering: The chapters on science and engineering offered a thorough and insightful examination of these developments' crucial role in achieving net zero emissions worldwide. The authors have skilfully analysed the many-faceted aspects of climate change mitigation to move towards a sustainable future, highlighting the necessity for cutting-edge technologies, sustainable behaviours, and interdisciplinary collaboration. The scientific underpinning offered in these chapters clarifies the intricate dynamics underlying climate change, providing a firm foundation for making wise decisions and formulating policies. The engineering solutions presented also provide viable approaches for lowering greenhouse gas emissions, raising energy efficiency, and improving renewable energy sources. Our comprehension of the problems we confront is improved by the seamless integration of science and engineering, and it also gives us hope by demonstrating the transformative potential of human ingenuity in achieving a net zero carbon future. Researchers, decision-makers, practitioners, and the public can all benefit from the chapters on science and engineering as they work together to create a sustainable and resilient world.

Part II Design and Innovation: The chapters on design and innovation make it clear how important it is for people to think creatively, plan strategically, and use cutting-edge innovation to move humanity toward a net zero future. The authors have thoroughly investigated the mutually beneficial relationship between design thinking and sustainability, demonstrating how deliberate design may simultaneously promote environmental advancement, economic prosperity, and societal well-being. Innovation and sustainability working together allow us to lessen the negative consequences of climate change and move us closer to a circular, regenerative economy. These chapters emphasise the significance of holistic and human-centred design strategies that give equal weight to social equality and environmental conservation. The case studies and best practices show how cutting-edge design approaches paired with environmentally friendly technologies may significantly lower carbon footprints, increase resource efficiency, and enhance the overall quality of life of people worldwide. Part II also emphasises the importance of interdisciplinary cooperation, promoting the fusion of concepts and knowledge from many fields. The adoption of sustainable design and innovation solutions, as well as the promotion of a culture of continuous innovation, depend on this collaboration. The chapters on design and innovation catalyse change, inspiring innovators, designers, policymakers, and citizens to engage in transformative actions that align to achieve net zero emissions. We can make significant advancements toward a more sustainable, equitable, and harmonious future for everyone by utilising the power of creativity and innovation.

Part III Energy Sector: The chapters on the energy sector show the significance and direct impact of the sector to achieve net zero. The authors have introduced and explained complex scientific methods, tools, techniques, technologies, and engineering involved in the energy sector, focusing on renewable energy, from essential photovoltaic technologies to more advanced and applied technologies such as transparent solar panels for windows in the built environment. Their fundamental research and experimental studies with different materials and proof-of-concept demonstrations show the sector's potential for further development and improvement, such as improved energy efficiency and performance of renewable energy technologies. The case studies revealed how researchers in the energy sector work across material science, chemistry, sustainability science, and engineering to successfully develop, test, evaluate, and disseminate new technologies, products, and systems in this sector. The authors have identified promising future research areas such as understanding the limits of different renewable energy system technologies and of infrastructure development, mechanisms, and technologies for recycling modules and devices at the end of their service lifetimes, and scaling up and adoption of pioneering technologies.

Part IV Transportation Sector: The chapters on the transportation industry offer a thorough analysis of one of the most critical and challenging fields in the worldwide effort to achieve net zero emissions. The authors have expertly negotiated the challenging terrain of the transportation industry, shedding light on its crucial contribution to environmental sustainability and the control of climate change. Through an analysis of the various aspects of this sector, they have highlighted the urgent need for efficient, sustainable, and equitable transportation systems. To revolutionise transportation, these chapters stress the need for a multidimensional strategy considering technical advancements, legislative changes, behavioural modifications, infrastructural improvements, and systemic transformation (Gota & Huizenga, 2023). They strongly emphasise the sector's ability to reduce its carbon footprint using electrification, hydrogen-based fuels, and other cutting-edge technology. The authors also underline the significance of sustainable urban planning and effective public transit networks to lessen individual dependency on fossil fuel-powered vehicles. Bergero et al. (2023) also argue that to achieve fossil fuel reductions, there is a need to replace fossil fuels with significant amounts of synthetic or biofuels with net zero emissions (sustainable fuels), which may be significantly more expensive. Part IV also promotes robust policy frameworks and international collaboration to bring about systemic transformation in the transportation industry. Implementing incentives, rules, and global agreements can be effective accelerators for the move toward a net zero carbon transportation sector.

Part V Fashion and Service Sectors: The chapters on the fashion and service industries examine two critical fields whose environmental influence needs to be more frequently reviewed. The authors have expertly highlighted the complicated interactions between the fashion and service industries and the urgent need for them to incorporate sustainability into their operations for a more sustainable future. Readers are confronted with a harsh understanding of the adverse effects of fast fashion on the environment and society in the chapters devoted to the fashion industry. The authors stress the need to transition to sustainable and circular fashion methods, promoting ethical production, responsible sourcing, and less waste. Additionally, cutting-edge technologies and material selections are crucial in modernising this industry and presenting alternative environmentally friendly materials and manufacturing techniques.

Similarly, the writers emphasise the significance of sustainable product-service system practices, energy efficiency, waste reduction, and responsible resource management in the chapters devoted to the service sector. These chapters discuss how companies in the service industry can lower their carbon footprint and help to achieve the net zero objective through strategic decisions and thoughtful choices, covering everything from sustainable event planning to green hospitality. The key message is that sustainability needs to be prioritised in both the fashion and service industries, influencing everything from product design to consumer behaviour and operational procedures. Manufacturing is quickly being surpassed by services to become the largest sector of the global economy (Randhawa & Scerri, 2015). These chapters strongly emphasise how consumer awareness and demand can drive different sectors of the economy toward environmentally friendly practices.

Summary

Considering what is discussed in its constituent chapters, this book is a crucial and necessary addition to our quest for a sustainable and environmentally conscientious future. The book emphasises the urgent need for a net zero carbon society and the critical role research leaders must play in directing this revolutionary paradigm shift. Readers are taken on a thorough investigation of ground-breaking research, innovative technologies, and forward-thinking strategies crucial to the fight against climate change. The book chapters present insight into the ground-breaking work of present and future research leaders to combat climate change, cut carbon emissions, and transform pre-existing mindsets. Throughout the book, the value of interdisciplinary collaboration is consistently demonstrated, highlighting the significance of a comprehensive strategy combining science, engineering, design, innovation, energy, transportation, fashion, and services. A holistic viewpoint is essential to address the complex problems caused by climate change and to create a sustainable and equitable society. The book also emphasises the necessity of promoting young researchers' creativity and supporting their commitment to sustainability. The book aims to ensure a constant push toward a net zero future by fostering and assisting the next generation of researchers. By doing this, we lay the road for a legacy of innovation with a long-lasting effect. In short, *Research Journeys to Net Zero: Current and Future Leaders* promotes action and provides information and education. It offers hope, indicating that with steadfast resolve, teamwork, and research-driven solutions, we can create a world in harmony with our environment, and ultimately progress towards a net zero carbon society.

References

Agrawal, R., Priyadarshinee, P., Kumar, A., Luthra, S., Garza-Reyes, J.A., & Kadyan, S. (2023). Are emerging technologies unlocking the potential of sustainable practices in the context of a net zero economy? An analysis of driving forces. *Environmental Science and Pollution Research International*, pp. 1–19.

Bergero, C., Gosnell, G., Gielen, D., Kand, S., Bazilian, M., & Davies, S.J. (2023). Pathways to net zero emissions from aviation. *Nature Sustainability*, 6, pp. 404–414. https://doi.org/10.1038/s41893-022-01046-9.

Franco, M., & Rodrigues, M. (2019). Sustainable practices in SMEs: Reducing the ecological footprint. *Journal of Business Strategy*, 42(2), pp. 137–142. https://doi.org/10.1108/JBS-07-2019-0136.

Fankhauser, S., Smith, S.M., Allen, M., Axelsson, K., Hale, T., Hepburn, C., Kendall, J. M., Khosla, R., Lezaun, J., Mitchell-Larson, E., Obersteiner, M., Rajamani, L., Rickaby, R., Seddon, N., & Wetzer, T. (2022). The meaning of net zero and how to get it right. *Nature Climate Change*, 12, pp. 15–21. https://doi.org/10.1038/s41558-021-01245-w.

Gota, S., & Huizenga, C. (2023). *The Contours of a Net Zero Emission Transport Sector in Asia – ADO 2023 Thematic Report*. Available at: https://www.adb.org/sites/default/files/institutional-document/874256/adotr2023bp-net-zero-emission-transport-asia.pdf.

Kitts, R.L., Christodoulou, J.N., & Goldman, S.J. (2011). Promoting interdisciplinary collaboration: Trainees addressing siloed medical education. *Academic Psychiatry*, 35, pp. 317–321.

Qader, A., Zhang, J., Ashraf, S.F., Syed, N., Omhand, K., & Nazir, M. (2022). Capabilities and opportunities: Linking knowledge management practices of textile-based SMEs on sustainable entrepreneurship and organizational performance in China. *Sustainability*, 14(4), pp. 2219. https://doi.org/10.3390/su14042219.

Randhawa, K., & Scerri, M. (2015). Service innovation: A review of the literature. In R. Agarwal et al. (eds.), *The Handbook of Service Innovation*. Springer-Verlag, pp. 27–51.

Sadowski, M., Perkins, L., & McGarvey, E. (2021). *Roadmap to Net Zero: Delivering Science-Based Targets in the Apparel Sector*. Working Paper. Washington, DC: World Resources Institute. Available at: https://doi.org/10.46830/wriwp.20.00004.

INDEX

Note: Locators in *italic* indicate figures.

Printed in the United States
by Baker & Taylor Publisher Services